AF678326

Safety Concerns for Herbal Drugs

Safety Concerns for Herbal Drugs

Divya Vohora
Hamdard University, New Delhi, India

S. B. Vohora
Hamdard University, New Delhi, India

CRC Press
Taylor & Francis Group
Boca Raton London New York

CRC Press is an imprint of the
Taylor & Francis Group, an **informa** business

CRC Press
Taylor & Francis Group
6000 Broken Sound Parkway NW, Suite 300
Boca Raton, FL 33487-2742

© 2016 by Taylor & Francis Group, LLC
CRC Press is an imprint of Taylor & Francis Group, an Informa business

No claim to original U.S. Government works

Printed on acid-free paper
Version Date: 20151008

International Standard Book Number-13: 978-1-4822-5661-1 (Hardback)

This book contains information obtained from authentic and highly regarded sources. Reasonable efforts have been made to publish reliable data and information, but the author and publisher cannot assume responsibility for the validity of all materials or the consequences of their use. The authors and publishers have attempted to trace the copyright holders of all material reproduced in this publication and apologize to copyright holders if permission to publish in this form has not been obtained. If any copyright material has not been acknowledged please write and let us know so we may rectify in any future reprint.

Except as permitted under U.S. Copyright Law, no part of this book may be reprinted, reproduced, transmitted, or utilized in any form by any electronic, mechanical, or other means, now known or hereafter invented, including photocopying, microfilming, and recording, or in any information storage or retrieval system, without written permission from the publishers.

For permission to photocopy or use material electronically from this work, please access www.copyright.com (http://www.copyright.com/) or contact the Copyright Clearance Center, Inc. (CCC), 222 Rosewood Drive, Danvers, MA 01923, 978-750-8400. CCC is a not-for-profit organization that provides licenses and registration for a variety of users. For organizations that have been granted a photocopy license by the CCC, a separate system of payment has been arranged.

Trademark Notice: Product or corporate names may be trademarks or registered trademarks, and are used only for identification and explanation without intent to infringe.

Visit the Taylor & Francis Web site at
http://www.taylorandfrancis.com

and the CRC Press Web site at
http://www.crcpress.com

Learn all that you can from ancient Indian systems of medicine, but do not believe that the last word could have been said thousands of years ago.

Jawaharlal Nehru

Contents

List of Tables

Foreword

There is increasing interest in the use of medicinal plants for therapeutic purposes in this country and worldwide. There are several reasons for this. The dawn of a realization that allopathy does not offer cures for chronic diseases like hypertension, diabetes, and bronchial asthma, whereas herbal products based on the experience and heritage of Ayurveda and the Unani System of Medicine could still offer hope. Also, increasing resistance of microbes to antibiotics due to their misuse may soon lead to a situation where effective antibiotics will cease to exist. Again, there may be very effective plants that possess antimicrobial properties.

Unfortunately, these optimistic possibilities are overshadowed by allopathic physicians who are concerned about the safety of herbal products. It is therefore appropriate and timely that at this juncture, Professor S.B. Vohora, an eminent professor of toxicology and a lifelong highly respected researcher on medicinal plants, together with Professor Divya Vohora, have written a book on the *Safety Concerns for Herbal Drugs*. The authors provide information so we can judge for ourselves the balance between the risks and benefits associated with the therapeutic use of medicinal plants. Led by World Health Organization (WHO) scientists and clinical investigators, regulators in Western countries have accepted the fact that when approving clinical trials for herbal products, efficacy is of less concern than safety.

What is however always a problem with drug regulatory agencies anywhere in the world is the use of herbomineral products many of which contain heavy metals. Scientists of the allopathic system of medicine and regulators are not willing to approve the use of such preparations unless a lack of toxicity has been unequivocally demonstrated. Another issue often discussed is whether herbal products, which have been used for hundreds of years, should be approved for clinical trials or even approved for use without undergoing preclinical toxicological tests. As scientists, the authors provide compelling reasons for treating all such issues on the basis of scientific principles. Unquestionably, this book will be widely read and discussed.

Several practitioners of the traditional systems of medicine and scientists in the developed world believe that herbal products do not induce toxic reactions because they are based on plants. For those of us who work in this field, we know that this is not so. It is to the credit of the authors that they have documented 134 instances of medicinal plants which are inducing side effects in humans. These data were collected from India and from 27 other countries. References are provided so that readers can follow up on their observations. In this book, the regulatory status for herbal products in 73 countries is reviewed. This information is very valuable since there is so much variation in regulations which makes it difficult for manufacturers and researchers. A fair degree of uniformity would help in a better and more rational use of herbal remedies.

Professors Divya Vohora and S.B. Vohora deserve our thanks for this very useful and informative book, which will undoubtedly help in the rational use of herbal products.

Professor Ranjit Roy Chaudhury
National Professor of Pharmacology Advisor
Department of Health and Family Welfare
Government of National Capital Territory of Delhi
India

Preface

There is no dearth of scientific literature on the benefits and importance of traditional medicinal systems and natural products (especially medicinal plants) in global healthcare programs in general, and developing countries, in particular. This is not yet another book on the subject. It stands out from the lot by focusing the attention of the users on their safety and toxicity aspects: danger from large-scale misuse and abuse, self-prescription, substitution, adulteration, concurrent use with modern medicines (without expert consultation), hazardous but avoidable drug interactions, risk groups, and present status of drug regulations. The regulatory laws and their enforcement are inadequate for more than 50% of member states of the World Health Organization. The spirit of inquiry and a dispassionate approach constitute the basis of scientific research. Exponents of the traditional systems of medicine and herbal drug manufacturers promote these drugs as totally safe. This book ventures into the relatively unexplored (or deliberately hidden) side of the picture. We present (1) a survey of approximately 1500 medicinal plants and herbal products, (2) 59 global (from 27 countries) and 75 Indian examples of toxic and adverse effects and drug interactions, and (3) the current status of regulatory laws and their enforcement in 73 countries to support this contention. There is an urgent need for a balanced approach to tackle the problems before they reach an alarming stage. Adamant or blind faith in ancient remedies has hampered the progress of traditional systems of medicine. Let them flourish to their full potential by keeping pace with the latest research and advancements. This book is a step toward that direction. Assistance from various organizations or persons is gratefully acknowledged.

Divya Vohora
Shashi B. Vohora
New Delhi, India

Acknowledgments

- Dr. Kin Shein, Traditional Medicine Program, World Health Organization: Regional Office for South-East Asia, New Delhi, India
- Dr. V.M. Katoch, Director General, Indian Council of Medical Research, New Delhi, India
- Professor Ranjit Roy Chaudhury, Chairman, Apollo Hospitals Education and Research Foundation, New Delhi, India
- Dr. Sheikh Raisuddin, Head, Department of Medical Elementology and Toxicology, Faculty of Science, Hamdard University, New Delhi, India
- Professor S.K. Kulkarni, Director, Bombay College of Pharmacy, Mumbai, India
- Dr. Zahoor A. Shah, Assistant Professor, College of Pharmacy and Pharmaceutical Sciences, University of Toledo, Ohio, United States
- Ms. Yashuro Sensho from the Japanese Embassy
- Mr. Slava Zeman from the Australian High Commission at New Delhi, India

Authors

Divya Vohora, M.Pharm., Ph.D., is a professor of pharmacology and in charge of the pharmaceutical medicine program for the Faculty of Pharmacy, Jamia Hamdard at New Delhi, India. She is also the assistant dean of Students' Welfare. She has more than 17 years of teaching and research experience and has published 90 publications (including 2 books, 5 book chapters, 21 reviews, and 62 research papers) in reputed national and international journals with more than 700 citations. Her major areas of interest are central nervous system (CNS) pharmacology (with an emphasis on epilepsy, neurodegenerative disorders, and cognitive function), osteoporosis, histamine, and clinical pharmacology. She has previously edited a multiauthor book with global participation on H3 receptors (CRC Press, 2009). Dr. Vohora has received grants and worked as a principal investigator for various research projects that were funded by the All India Council for Technical Education (AICTE), the Indian Council of Medical Research, the Council for Scientific and Industrial Research (CSIR), the University Grants Commission (UGC), and the Department of Science and Technology (DST), Government of India. She is also coordinating the University Grants Commission Special Assistance Program of the Department of Pharmacology at Department Research Support (DRS) Phase II level. Under her supervision, 32 postgraduate (24 M.Pharm. and 7 Ph.D.) degrees have been awarded. She has participated in many national and international conferences and symposia and presented invited lecture and research papers in India and abroad (including Australia, France, London, Malaysia, Poland, Singapore). Dr. Vohora was an organizing secretary, a co-organizer, and a chairperson for some of these meetings and workshops. She has received many prizes and awards in recognition of her scientific contributions, some of which include: the Chandra Kanta Dandiya Prize for best published paper in pharmacology, the DST Fast Track Award for Young Scientists, and the AICTE Career Award for Young Teachers. She is a member of many learned societies, professional bodies, and expert committees including the International Brain Research Organization (France), the Committee for National Societies (CNS), the International Osteoporosis Foundation, the Indian Pharmacological Society, and the Indian Society for Bone and Mineral Research. Dr. Vohora has also served on editorial boards and has been a reviewer for more than 30 journals of repute.

Shashi B. Vohora, M.V.Sc., Ph.D., D.Sc., retired as a professor and head of the Department of Medical Elementology and Toxicology, Hamdard University, New Delhi. Earlier, he worked for the Indian Council of Medical Research, at the Post-Graduate Institute of Medical Education and Research, Chandigarh, Lady Hardinge Medical College, New Delhi; and later for the Institute of History of Medicine and Medical Research, New Delhi. He was an honorary project officer in the Drug Standardization Research Unit of the Central Council for Research in Unani Medicine at New Delhi. He has over 30 years of teaching, research, and editing experience with more than 200 research publications, including 18 books. His papers have been cited in books published in the United States, and some of his publications have been translated into Polish, Italian, and Hungarian. He is the first to hold a Ph.D. in science policy in India. His major areas of interest include neuropsychobehavioral studies, medicinal plants, medical elementology, and traditional mineral drugs. Dr. Vohora was awarded three research projects: two by the Indian Council of Medical Research and one by the Central Council for Research in Unani Medicine on neuropsychopharmacological, immunomodulatory studies, and hypoperfusion- and reperfusion-induced brain injury. Two M.Pharm. (1991, 1992) and nine Ph.D. (1993–2006) degrees were awarded research under his supervision. One candidate carried out postdoctoral research under the guidance of a Research Associateship from the Council of Scientific and Industrial Research (2001–2003). Dr. Vohora was the organizing secretary of the IHMMR- and WHO-sponsored First International Conference on Elements in Health and Disease, New Delhi (1983), the National Symposium on the Development of Indigenous Drugs in India during the last 25 years, New Delhi (1988), and convener of the National Symposium on Toxicology and Environmental Health, New Delhi (1998). He has been on scientific visits and academic tours to Poland, Hungary, Germany, France, Switzerland, Italy, Ceylon, Pakistan, and Nepal, and has delivered several invited and plenary lectures in India and abroad. He is a nominated member of the Board of Governors, Anstad University, Virgin Islands; a founding member and promoter for the International Union of Elementologists; a nominated member of the DST Expert Panel on *Bhasmas*; and a member of the International Brain Research Organization (France), the International Society for Trace Element Research in Humans, the Indian Pharmacological Society, the Society of Biosciences (India); and a Correspondent Member, Curare (Germany), and Investigational Drugs (United Kingdom).

1 Introduction

1.1 HERBAL DRUGS: WIDE USAGE AND POPULARITY

Plants are the original drugs. The Sanskrit word *Oushadhi* denotes the plant kingdom (flora), and the French word *drogue* means a dry herb. The earliest mention of medicinal use of plants is found in the *Rig Veda* (3000 BC), which is one of the oldest repositories of human knowledge (Vohora 1979). There are three kinds of herbal medicines: raw plant materials, processed plant materials, and medicinal herbal products (WHO 1998). The World Health Organization (WHO) defines herbs to include plant material such as leaves, flowers, fruit, seeds, stems, wood, bark, roots, rhizomes, or other plant parts, which may be whole, fragmented, or powdered (WHO 2000), and are used for their medicinal or cosmetic properties. Vohora et al. (1973) reviewed common Indian vegetables, which form part of the daily meals of the masses, to show that these edible plants are attributed with many medicinal properties. Many modern drugs originated from plants, and natural products constitute a major source of drug development in the pharmaceutical industry (Harvey 2008). Out of 800,000 plant species on Earth, about 25% have been categorized, and only a small fraction has been pharmacologically evaluated. Despite this fact, the use of herbal medicines and nutraceuticals continues to expand rapidly across the world in different healthcare settings (WHO 2004; Abdin and Abrol 2006). A tremendous increase in the use of herbal products has been observed during the last three decades with not less than 80% of the world population relying on them for primary healthcare. Bandaranayake (2006) attributed this to an increased tendency for self-medication. Ekor (2013) has listed varied factors which are responsible for increased patronage of herbal drugs including dissatisfaction with orthodox pharmaceuticals; the erroneous belief that herbal products are superior to manufactured medicines; greater rapport with traditional healers; patients' belief that herbal medicines might be effective in certain diseases for which modern medicines have been ineffective, inadequate, and expensive; and improvement in quality and efficacy of herbal drugs following utilization of modern science and technology in their development. Over the past decade, the surge in herbal medicines has been estimated to be 40% of all healthcare services in China. The percentage of the population that has used herbal medicines at least once is Australia (48%), Canada (70%), United States (42%), Belgium (38%), and France (75%) (Ekor 2013). This shows that the usage is not restricted to developing countries in Asia and Africa.

1.2 CAUSES FOR ALIENATION FROM MODERN MEDICINE

Side effects and the high costs of treatment have been the two major factors which are turning people away from modern medicines. Corticosteroids and antibiotics are attributed with serious toxic effects and adverse reactions. In the United States, about nine million drug reaction cases occur annually, out of which about 30,000 cases are reported to be fatal (Abdin and Abrol 2006). It is widely believed and repeatedly stated that herbal drugs, being "natural," are devoid of all toxic and adverse side effects and reactions. Medicinal plants and their formulations are promoted as totally safe by exponents of traditional systems of medicine and herbal drug manufacturers. Consumers mostly self-prescribe these drugs. Often patients do not consider it important to inform their family physician about their intake, and continue to take "harmless" herbal products simultaneously with prescribed medicines. This is a cause for great concern as it might lead to serious (but avoidable) drug interactions (Hailer et al. 2002). A national survey in the United States (1990–1997) revealed that only 40% of the people who used herbal remedies informed their primary healthcare provider about it (Einsberg et al. 1998).

1.3 QUESTIONS AND AIMS OF THE STUDY

- Are herbal drugs totally devoid of adverse effects (1) when used alone, (2) as herbal formulations, and (3) in concurrent use with modern medicines?
- Is their use safe when the patient is under stress and receiving modern medicines in the following conditions: (1) pregnancy, (2) pediatric and geriatric population, (3) preoperative and postoperative use, and (4) in patients with diabetes, hepatic, cardiac, and neuropsychiatric disorders?

This study addresses the above-mentioned questions and aims to probe the safety aspects of herbal drugs with a dispassionate approach. The following methods were used for investigation:

a. A global survey of the literature with a special emphasis on India
b. Mass-scale requests (for input) to various scientific organizations; funding agencies and national and regional institutes; universities; drug regulatory authorities; prominent herbal drug manufacturers; eminent scientists; friends and colleagues; research scholars, etc.

REFERENCES

Abdin, M.Z. and Abrol, Y.P. (eds.) 2006. *Traditional System of Medicine*. Narosa Publishing House, New Delhi.

Bandaranayake, W.M. 2006. Quality control, screening, toxicity and regulation of herbal drugs. In: Ahmad, I., Aqil, F., and Owais, M. (eds.) *Modern Phytomedicine. Turning Medicinal Plants into Drugs*. Wiley-VCH Verlag GmbH & Co., Weinheim.

Einsberg, D.M., Davis, R.B., and Ernst, E. 1998. Trends in alternative medicine use in the United States 1990–1997: Results of a follow up national survey. *JAMA* 280: 1569–1575.

Ekor, M. 2013. The growing use of herbal medicines: Issues relating to adverse reactions and challenges in monitoring safety. *Frontiers in Pharmacol* 4: 1–10.

Hailer, C.A., Dyer, J.E., Ko, R.J., and Olson, K.R. 2002. Making a diagnosis of herbal-related toxic hepatitis. *West J Med* 176: 39–44.

Harvey, A.L. 2008. Natural products in drug discovery. *Drug Discovery Today* 13: 894–901.

Vohora, S.B. 1979. Research on Medicinal Plant in India: Effort and Achievements (1947–1976). Thesis PhD (Science Policy), Jawaharalal Nehru University, New Delhi.

Vohora, S.B., Rizwan, M., and Khan, J.A. 1973. Medicinal uses of common Indian vegetables. *Planta Medica* 23: 381–393.

WHO. 1998. *Guidelines on Appropriate Use of Herbal Medicines*. WHO regional publication, Western Pacific Series, no. 23, Manila.

WHO. 2000. *General Guidelines for Methodologies on Research and Evaluation of Traditional Medicine*. World Health Organization, Geneva.

WHO. 2004. *Guidelines on Safety Monitoring of Herbal Medicines in Pharmacovigilance Systems*. World Health Organization, Geneva.

2 Toxicity and Adverse Effects of Herbal Drugs

2.1 INTRODUCTION

The entry of counterfeit, poor quality, or adulterated herbal products in the international market, is driven largely by the strong revenue that it can generate, which is a cause of serious safety concern (Verma 2013). Despite the safety claims by the exponents of traditional systems of medicine, considerable information is available on the adverse effects of herbal drugs (McGregor et al. 1989; Desmet et al. 1992; Ernst 2000). The potential toxicity of herbal drugs is not new. In several countries where herbal drugs are commonly used by the masses, it is well known that some plants must be used with caution because adverse effects may be caused inherently by toxic herbal overdoses, drug-to-drug interactions, and other factors (Bnouham et al. 2006).

2.2 INCIDENCE

A pilot study was carried out by the National Poison Unit, Gay's and St. Thomas Hospital Trust, London to probe the frequency and severity of reports on the adverse effects of traditional medicines and food supplements. Emergency case inquiries to the unit, surveyed retrospectively (during 1983–1988) and prospectively (in 1991), showed a total of 5563 inquiries, with 657 patients (12%) presenting with symptoms at the time of initial inquiry. Links between exposure and clinical effects were found in 49 cases. Most of the cases were attributed to heavy metal exposure from Asian medicines through metal content, as a constituent of such preparations or due to contamination or adulteration of herbal formulations. The symptoms included gastrointestinal, hepatotoxic, dermatological, respiratory, and central nervous system problems. The effects, identified by the survey, were considered serious enough to cause concern and necessitated continued surveillance for health risk assessment (Perharic-Walton and Murray 1992). Increased incidence was observed in cases that were reported to the Malaysian Adverse Drugs Reaction Advisory Committee; the number of cases rose from 11 in 1997 to 23 in 1999. Approximately 37% of the 5000 cases of renal problems in Malaysia were attributed to the chronic use of traditional herbal preparations (Hussin 2001). The misconception that natural products are non-toxic and devoid of adverse effects is not limited to developing countries. According to a recent report, it also exists in highly developed countries where the general population resorts to natural products without proper awareness or information on the associated risks, particularly in the event of chronic or excessive use (UNESCO 2013).

Raynor et al. (2011) warned buyers of five commonly used herbal products (St. John's Wort, Asian Ginseng, *Echinacea*, Garlic, and Ginkgo) where the information provided on herbal product labels, provided on herbal product labels, marketed over-the-counter (OTC) and online in Europe, North America, and Australia, may be misleading. Analysis revealed that overall, 51 out of 68 (75%) contained none of the key safety messages which are required in the monographs of the U.S. National Center for Complementary and Traditional Medicine. This included 4 out of 12 for St. John's Wort, 2 out of 12 for Ginkgo, and 6 out of 7 for Asian Ginseng products. Only two of the purchased products, registered under the new European Union regulations (for St. John's Wort) contained 85% of the mandatory safety messages.

Online sources of herbal product information are grossly inadequate and disturbing. Owens et al. (2014) conducted online research for 13 common herbal drugs (e.g., Black Cohosh, *Echinacea*, Garlic, Ginkgo, Ginseng, Green Tea, Kava, Saw Palmetto, St. John's Wort, etc.). They examined 1179 websites for clinical claims, warnings, and other safety information. They found that (1) less than 8% of retail sites provided information regarding potential adverse effects and drug interactions, (2) only 10.5% of the sites advised consultation with healthcare personnel, and (3) less than 3% cited scientific literature in support of their claims. While certain nonretail sites may be more reliable, healthcare professionals and physicians should be more aware of the variable quality of such products in order to make informed decisions and help their patients. Self-prescription and underreporting are widely prevalent. Therefore, a large proportion of toxicity reactions to herbal drugs may go unrecognized (Hailer et al. 2002).

2.3 CLASSIFICATION AND FACTORS AFFECTING HERBAL TOXICITY

Vohora et al. (1973), Degraeve (1978), Vohora (1979), Hussin (2001), Premila (1989), Bnouham et al. (2006), and Faghihi and Radan (2011).

2.3.1 Endogenous Toxicity Due to the Inherent Chemical Composition of Herbal Drugs

Even the harmless potato may exhibit toxic effects under certain conditions. When it sticks out of the soil, it turns green due to the formation of the toxic alkaloid *solanine*. The latter is found in immature potatoes and in growing shoots. Fortunately, the toxic effects are mild due to (1) low alkaloid content and (2) peeling of potatoes and removing the skin and shoots before eating the parts where the toxic alkaloid is concentrated.

Celery is known to contain *psoralens*, a group of substances that cause toxic dermal reaction up on exposure to ultraviolet rays. A report from the Ministry of Health, Malaysia stated that almost 95% of the unregistered traditional herbal preparations on the local market contained steroids. Some other examples of endogenous toxicity inlcude alkaloids in *Datura stramonium, Aconitum chasmanthus*, and *Strychnos nux vomica;* glycosides in *Calotrapis gigantea* and *Nerium indicum*; and phytotoxins in the seeds of *Abrus precatorius* and *Ricinus communis.* Experimental studies in rats have shown that chronic exposure to β-Asarone from the oil of *Acorus calamus* (rhizomes) was responsible for causation of malignant tumors in the duodenum. Therefore, calamus as a food additive was banned in the United States.

2.3.2 Exogenous Toxicity Due to External Factors

a. Contamination: Microbial, chemical, heavy metal, radioactive material, and so on.
b. Pesticide residue
c. Faulty identification and processing
d. Substitution with cheaper drugs or adulteration
e. Inadequate storage and transport conditions, and so on

2.3.3 Mixed Toxicity: Hybrids of Endogenous and Exogenous Factors

Polypharmacy is the rule (rather than the exception) in herbal drug therapy. Formulations usually contain many herbal and/or mineral ingredients with synergistic and antagonistic effects. This may cause mild-to-serious drug interactions (herb + herb or herb + modern drug). The claims that the ingredients act as correctives (with a reduction of harmful effects and potentiation of efficacy) have not been scientifically verified (or refuted) for most of the herbal formulations.

2.4 RISK GROUPS

Certain population groups should be extra cautious as they are more susceptible to herbal adverse reactions and toxicity (MacGregor et al. 1989; Ang-Lee et al. 2001; Hussin 2001; Hailer et al. 2002):

a. Pregnant and nursing women
b. Infants and children
c. Elderly persons
d. Patients with diabetes, cardiovascular, hepatic, respiratory, renal, and neuropsychiatric disorders
e. Preoperative and postoperative cases who are required to take other life-sustaining drugs (these groups are highly sensitive and may show exaggerated toxic effects or drug reactions)

REFERENCES

Ang-Lee, M.K., Moss, J., and Yuan, C.S. 2001. Herbal medicines and perioperative care. *JAMA* 286: 208–216.

Bnouham, M., Mehrfour, F.Z., Elachoui. M., Legssyer, A., Mekhfi, H., Lamnouer, D., and Ziyyat, A. 2006. Toxic effects of some plants used in Moroccan traditional medicine. *Moroccan J Biol* 2–3: 21–30.

Degraeve, N. 1978. Genetic and related effects of *Vinca rosea* alkaloids. *Mutation Res* 55: 31–42.

Desmet, P.A.G., Kelle, K., Hansel, R., and Chandler, R.P. 1992. *Adverse Effects of Herbal Drugs*. Springer Verlag, Berlin, p. 264.

Ernst, E. 2000. Adverse effects of herbal drugs in dermatology. *British J Dermatol* 143: 923–929.

Faghihi, G. and Radan, M. 2011. Side effects of herbal drugs in dermatology. *J Cosmetics, Dermatol Sci App* 1: 1–3.

Hailer, C.A., Dyer, J.E., Ko, R.J., and Olson, K.R. 2002. Making a diagnosis of herbal-related toxic hepatitis. *West J Med* 176: 39–44.

Hussin, A.H. 2001. Adverse effects of herbs and drug-herbal interactions. *Malaysian J Pharmacy* 1: 39–44.

MacGregor, F.B., Abernethy, V.E., Dahabra, S., Cobden, I., and Hayes, P.C. 1989. Hepatatoxicity of herbal remedies. *BMJ* 299: 1156–1157.

Owens, C., Baergen, R., and Puckett, D. 2014. Online sources of herbal product information. *Am J Med* 127: 109–115.

Perharic-Walton, L. and Murray, V. 1992. Toxicity of Chinese herbal remedies. *Lancet* 340: 674.

Premila, M.S. 1989. Safety of herbal drugs, the need for caution. In: *Research and Development of Indigenous Drugs* (eds. Dandiya, P.C. and Vohora, S.B.), Institute of History of Medicine and Medical Research, New Delhi.

Raynor, D.K., Dickinson, R., Knapp, P., Long, A.F., and Nicolson, D.J. 2011. Buyer beware? Does the information provided with herbal products available over the counter enable safe use? *BMC Med* 9: 94–101.

UNESCO. 2013. *Report of the International Bioethics Committee on Traditional Medicine Systems and Their Ethical Implications. SHS/EGC/IBC-* 19/12/3 Rev. Paris, February 8.

Verma, N. 2013. Herbal medicines: Regulation and practice in Europe, United States and India. *Int J Herbal Med* 1: 1–5.

Vohora, S.B. 1979. Research on Medicinal Plant in India: Effort and Achievements (1947–1976). Thesis Ph.D. (Science Policy), Jawaharalal Nehru University, New Delhi.

Vohora, S.B., Rizwan, M., and Khan, J.A. 1973. Medicinal uses of common Indian vegetables. *Planta Medica* 23: 381–393.

3 The Global Picture

Examples of adverse effects of herbal drugs reported from Argentina, Australia, Austria, Bangladesh, Belgium, Brazil, Canada, China, France, Germany, Ghana, Hong Kong, Iran, Italy, Japan, Malaysia, Mauritius, Morocco, Netherlands, Nigeria, Portugal, South Africa, Spain, Sri Lanka, Turkey, United Kingdom, and the United States are presented in Table 3.1.

TABLE 3.1
Global Picture: 59 Examples of the Adverse Effects of Single-Component and Multicomponent Herbal Drugs from 27 Countries

S. No.	Country	Toxic and Adverse Side Effects or Interactions	Toxic or Active Principle	References
1.	Argentina	1. *Aristolochia struckeretii* exhibited high neurotoxic effects against rat hippocampal cells in culture. The model used embryonic rat cells to investigate the action on the development of neurons *in vitro* in stereotypical sequence that mimics their development *in vivo*. Neuron viability was also evaluated using a lactate dehydrogenase (LDH) assay. Researchers from the Ministry of Science and Technology, Colorado, and the Laboratorie de Neurobiologia Cellular y Molecular Instituto Investigacion Medica Mercedes y Martin Ferreyra claimed that such neurotoxic effects were reported for the first time in this plant and recommended intensive studies to elucidate the mechanisms that are involved.	Not identified (crude extract used)	Landau and Sanchez 2010
2.	Australia	2. *Zingiber officinale* (ginger) tea, used as a common remedy for morning sickness, was tested for its effects on pregnancy and fetal development in Sprague-Dawley rats. Daily intake (20–50 g L^{-1} from days 6–15 of gestation) revealed the following: a. No maternal toxicity. b. Significant embryonic loss in the treated group (double versus control rats). c. That while no gross morphological abnormalities were detected, fetuses in the treated group were found to be significantly heavier than control group fetuses. The study, carried out at the Charles Stuart University, Wagga Wagga, shows that *in utero* exposure to ginger tea results in increased embryonic loss with increased growth in surviving fetuses.	Not identified (tea used)	Wilkinson 2000

(*Continued*)

TABLE 3.1 (CONTINUED)
Global Picture: 59 Examples of the Adverse Effects of Single-Component and Multicomponent Herbal Drugs from 27 Countries

S. No.	Country	Toxic and Adverse Side Effects or Interactions	Toxic or Active Principle	References
		3. Uddin et al. from Griffith University, Queensland, investigated the cytotoxic activity of 32 extracts from 16 Bangladeshi plants against the following: a. Healthy mouse fibroblasts. b. Human cancer cell lines: Gastric (AGS), colon (HT 29), and breast (MDA-MB-4355) using (3- [4, 5-dimethyl thiozol-2-yl]-2, 5-diphentetrazolium bromide) colorometric assay. c. *Plants*: (1) *Acrostichum aureum* (Tiger Fern), (2) *Adiantum caudatum* (Mayur Shikha), (3) *Aegiceras corniculatum* (Khalsi), (4) *Ammania baccifera* (Jangli Mehndi), (5) *Argemone maxicana* (Shialkota), (6) *Blumea lacera* (Kukur Sunga), (7) *Bruguiera gymnorrhiza* (Kankra), (8) *Clerodendron remi* (Bon Jui), (9) *Cyanometra ramiflora* (Kucha), (10) *Ficus religiosa* (Pan Bot, Pipal), (11) *Hibiscus tiliaceous* (Bhola), (12) *Hygrophila auriculata* (Talmakhana), (13) *Limnophila indica* (Karpur), (14) *Mollugo pentaphylla* (Khet Papra), (15) *Pandanus foetidus* (Kewa Kata), and (16) *Xylocarpus moluccensii* (Passur). d. *Results*: The IC_{50} values (50% inhibition of cell growth) for most extracts ranged between 1 and 2 mg mL^{-1}. The authors opined that these values, though somewhat high, still point subtly toward selective activity. High IC_{50} values were attributed to a very low concentration of compounds of interest, which were considerably enriched after bioactivity-guided fractions. The findings were graded into different categories that are summarized as follows: i. *Low or no cytotoxic activity*: *9 out of 16 aqueous extracts and 3 out of 16 methanolic extracts* were placed under this category. The low cytotoxicity of aqueous extracts is of great significance for their traditional use for the treatment of various disorders other than cancer.	Not identified (crude extracts used)	Uddin et al. 2011

(*Continued*)

TABLE 3.1 (CONTINUED)
Global Picture: 59 Examples of the Adverse Effects of Single-Component and Multicomponent Herbal Drugs from 27 Countries

S. No.	Country	Toxic and Adverse Side Effects or Interactions	Toxic or Active Principle	References
2.	Australia	ii. *Selective cancer-cell cytotoxic activity*: Criteria: Low cytotoxicity (IC_{50} values: >2.5 mg mL^{-1}) against mouse fibroblasts and selective cytotoxicity (IC_{50} values: 0.2–2.3 mg mL^{-1}) against human cancer cell lines. Eleven plant extracts—seven methanolic extracts (plants 1, 4, 5, 8, 9, 13, and 16) and four aqueous extracts (plants 3, 7, 12, and 16)—were placed under this category. These plant extracts have great potential for the development of anticancer drugs. iii. *High nonselective cytotoxic activity*: The methanolic extract of plant *Blumea lacera* leaves exhibited the highest cytotoxic activity (IC_{50} values: 0.01–0.08 mg mL^{-1}) against all cell lines among the 32 extracts that were tested. Moderate cytotoxicity (IC_{50} values: 1.23–1.88 mg mL^{-1}) was observed with the methanolic extracts of *Adiantum caudatum*. Plant extracts, placed under this category, are a cause for safety concern for human use and warrant a probe by drug regulatory authorities.		
3.	Austria	4. Kramar and Kaiser from the Institute of Medicinal Chemistry, University of Vienna, reported the hepatotoxic effects of Imperotonin (a pure coumarin from the plants *Ammi majus* and *Imperatoria ostruthium*), with cirrhotic changes in rat liver, loss of oxidized nucleotides from liver, and inhibition of succinate-linked respiration and phosphorylation in hepatic mitochondria. They investigated the effect of this toxic principle from *A. majus* on energy transfer in mitochondria, Imperotonin (3–10^{-4} M)-inhibited respiration, and phosphorylation by 33% and 50%, respectively. Energy-linked transhydrogenase reactions were also significantly inhibited.	Imperotonin	Kramar and Kaiser 1968

(*Continued*)

TABLE 3.1 (CONTINUED)
Global Picture: 59 Examples of the Adverse Effects of Single-Component and Multicomponent Herbal Drugs from 27 Countries

S. No.	Country	Toxic and Adverse Side Effects or Interactions	Toxic or Active Principle	References
		5. Complementary and alternative medicine (CAM) is often used by patients with malignant glioma. Although several interactions of various CAM medicines with chemotherapy are known, a case report of severe liver toxicity with jaundice during radiotherapy with Temozolomide was first described by Melchardt et al. in 2014. The researchers attributed it to an interaction with a popular Chinese herbal formula following surgery for glioblastoma. After cessation of the herbal formula as well as chemotherapy, the liver enzymes slowly normalized. Due to tumor progression, the patient was retreated with Temozolomide for five cycles without toxicity.	–	Melchardt et al. 2014
4.	Bangladesh	6. Khan and Islam from South Eastern University, Chittagong, reported that the ethanolic extracts of four plants (250–500 mg kg^{-1}, po) exhibited significant analgesic activity against acetic acid-induced writhing in mice and cytotoxic effects in brine shrimp lethality assay with IC_{50} and IC_{90} values (μg mL^{-1}) ranging between 15.16–28.16 and 22.02–54.51, respectively. Acute toxicity studies (10–1000 mg kg^{-1}) revealed LD_{50} (mean ± S.D.,%) values as follows: *Acorus calamus* roots: 61.25 ± 26.66, *Kigella pinnata* bark: 52.50 ± 24.37, *Mangifera indica* bark: 58.70 ± 25.22, and *Tabernaemontana divericata* leaves: 56.25 ± 22.88. While these values are quite high, the authors considered that these plant extracts can still be preferred for the treatment of painful conditions and tumors.	Not identified (crude extracts used)	Khan and Islam 2012

(Continued)

TABLE 3.1 (CONTINUED)
Global Picture: 59 Examples of the Adverse Effects of Single-Component and Multicomponent Herbal Drugs from 27 Countries

S. No.	Country	Toxic and Adverse Side Effects or Interactions	Toxic or Active Principle	References
5.	Belgium	7. Degraeve from the Genetics Laboratory of the University of Liege critically reviewed the genetic effects of alkaloids of *Vinca rosea* (antitumor properties, effects on DNA and RNA synthesis, chromosomes, gametogenesis, mutagenicity, and teratogenicity). Vinblastine inhibited DNA synthesis *in uteri* of pregnant mice and exhibited teratogenic effects in hamsters. No chromosomal aberrations were observed in most experimental studies. There was no demonstrable mutagenicity in microorganisms and mammals, not very reassuring for the latter. The authors opined that since *Vinca* alkaloids are used exclusively for the treatment of neoplastic diseases, the risks (genetic and teratogenic) are surely less important versus other therapies (radiotherapy and alkylating agents) for such conditions.	Vincristine, Vinblastine	Degraeve 1978
		8. Creemers et al. from Klinisch Laboratorium, Baheiden, reported two cases of Belgian women with lead poisoning. The diagnosis was based on (1) clinical symptoms (abdominal pain), (2) anemia (normocytic, normochromic), (3) basophilic stippling of erythrocytes, and (4) elevated blood Pb levels. Upon further questioning, the patients admitted taking Ayurvedic pills (orange-red in color, composition not known). Chemical analysis of these pills revealed a remarkably high amount of Pb (31 mg per pill). The authors stated that while cases of Pb poisoning from the use of Ayurvedic formulations are emerging around the world, these are the first reports from Belgium.	Lead	Creemers et al. 2007

(Continued)

TABLE 3.1 (CONTINUED)
Global Picture: 59 Examples of the Adverse Effects of Single-Component and Multicomponent Herbal Drugs from 27 Countries

S. No.	Country	Toxic and Adverse Side Effects or Interactions	Toxic or Active Principle	References
		9. Coopman et al. from the Department of Analytical Toxicology, Chemiphar NV, Bruges, Belgium, reported a case of suicidal death after injection (intravenous or subcutaneous) of a castor bean (*Ricinus communis*) extract in a 49-year-old man. He was brought to the emergency department 24 hours after injecting himself. On admission, he was conscious and presented with a history of nausea, vomiting, diarrhea, dyspnea, vertigo, and muscular pain. Despite symptomatic and supportive care, he died after 9 hours due to multiple organ failure. Exposure was confirmed by identification of the biomarker Ricinine in blood, urine, vitreous humor using solid-phase extraction, liquid chromatography and tandem mass spectrometry. Based on clinical symptoms and results of toxicological analysis, the cause of death was attributed to the plant toxin.	Ricinine (Piperidine alkaloidal toxin)	Coopman et al. 2009
6.	Brazil	10. Rates from the School of Pharmacy, Federal University of Rio Grande do Sul Porto Alegre, reviewed various aspects of drug development from higher plants. He discussed the main strategies for obtaining drugs from natural sources, traditional beliefs, fields of knowledge involved, difficulties, and perspectives with the focus on the situation in Brazil regarding usage, trade, research, and regulations of therapeutic resources of natural origin. Generally, a lack of awareness of the potentially toxic plants used in folk medicine was highlighted with many examples:	a. Toxic lectins in *Plura creptans*, *Ricinus communis*, and *Abrus precatorius* b. Pyrrolizidine alkaloids in Comfrey, *Senecio bravilensis*, and *Lantana camara*	Rates 2001

(*Continued*)

TABLE 3.1 (CONTINUED)

Global Picture: 59 Examples of the Adverse Effects of Single-Component and Multicomponent Herbal Drugs from 27 Countries

S. No.	Country	Toxic and Adverse Side Effects or Interactions	Toxic or Active Principle	References
6.	Brazil	a. *Toxic accidents due to wrong botanical identification:* (1) *Coffee substitutes*: In Brazil, mainly in the North and Northeastern regions, it is common to use certain plants (bark, root, or seeds) to prepare infusions that are used as a substitute for coffee (*Coffea arabica*). These include *Caeseria sylvestris* (Cha-de-burge or Café-do-diabo), *Cardia coffeoides* (Café-de-mato), *Mucuna pluriocostato* (Café-dos-novegartes), and *Cassia occidentalis* (Fedegoso) resulting in intoxication. A number of deaths were reported from the consumption of *Hura crepitans* bark, another coffee substitute, south of Pará. It contains a toxic lectin, also found in other plants, for example, *Ricinus communis* (Marmona) and *Abrus precatorius* (Jequiriti). (2) The *wrong use of the plant Quebra-Pedra used as a diuretic and for treating gallstone problems*: The correct plant is *Phyllanthus niruri,* which is commonly confused and substituted with some plants from the genus *Euphorbia*; the latter is toxic. b. *Intoxication by common remedies:* Popular remedies made without legal authorization and sold by herbalists or even prescribed by religious leaders for use in rituals are quite rampant and have often resulted in toxic symptoms immediately after ingestion or later. (1) *Symphytum officinalis* (Confreti) as panacea, (2) *Aloe* spp. (Babosa), and (3) *Euphorbia tirucalli* (Aveloz) for treating cancer are very common. (4) Comfrey and (5) *Lantana camara* are now legally prohibited for internal use as these contain potentially hepatotoxic pyrozolidine alkaloids. (6) *Senecio brasilensis* is also rich in toxic pyrozolidine alkaloids. *Aloe* spp. latex may cause severe diarrhea.	c. Phorbol esters in latex of *Aloe* spp. d. Cardiac glycosides in *Nerium oleander* and *Thevetia peruviana* e. Coumarins f. Estrogenic compounds g. Tyramine h. Allergenic substances i. Sesquiterpene lactones j. Photosensitive compounds (Furocoumarins, Hypericin) k. Toxic alkaloids: Atropine, Scopolamine l. Calcium oxalate m. Mycotoxins, and so on	

(*Continued*)

TABLE 3.1 (CONTINUED)
Global Picture: 59 Examples of the Adverse Effects of Single-Component and Multicomponent Herbal Drugs from 27 Countries

S. No.	Country	Toxic and Adverse Side Effects or Interactions	Toxic or Active Principle	References
		c. *Cardiotonic plants*: (1) *Nerium oleander* and (2) *Thevetia peruviana*, rich in cardiac glycosides, have been reported to cause toxic accidents in domestic animals and human subjects (especially children). Some decorative plants, (3) *Calotropis procera* and (4) *Gomphocarpus fruticoses*, also caused such accidental poisoning.		
		d. *Toxic interactions with conventional medicines*:		
		i. *Plants containing coumarin derivatives*: These compounds may lead to hemorrhages by their own chronic use or through enhancing the effects of oral anticoagulants (e.g., Dicoumarol, Sodium Coumarins). Among the coumarine-rich plants widely used in folk medicine as flavoring agents are (1) *Mykonia* spp. (Guaco), (2) *Meliolotus officinalis* (Trevo-Doce-Amaric), and *Dypterix odorata* (Java Tonka).		
		ii. *Plants containing estrogenic compounds*: (1) *Panax* spp. (Ginseng) and (2) *Dioscorea* spp. (Inhame) can cause estrogenic effects per se and through synergistic action with steroid drugs. Such combinations should be avoided.		
		iii. *Plants with high tyramine content*: Tyramine is a phenylethylamine that is found in yeast products (e.g., cheese and wine), which can cause hypertensive accidents in patients treated with monoamine oxidase (MAO)-inhibitor drugs. Such plants are (1) Mushrooms, (2) *Portulacca* spp. (Onze-Horas), (3) *Phoradendron* spp., and (4) *Psittacanthus* spp. (Erva-de-Paeserhino), and so on.		
		iv. *Plants causing allergic problems or irritation*: Allergic reactions caused by plants, through contact with pollens, secretions, or volatile substances, are not uncommon. Irritation or dermatitis can occur from (1) *Urtica prenas* and other plants of the Urticaceae family; (2) *Croton* spp., (3) *Jater* spp., and (4) *Cnidoscolus* spp. from the family Euphorbiaceae; (5) *Mucuna pruriens*, (6) *Matricaria chamomilla* (Chamomile), (7) *Arnica montana* (Arnica), (8) *Pffia* spp., and so on.		

(Continued)

TABLE 3.1 (CONTINUED)
Global Picture: 59 Examples of the Adverse Effects of Single-Component and Multicomponent Herbal Drugs from 27 Countries

S. No.	Country	Toxic and Adverse Side Effects or Interactions	Toxic or Active Principle	References
6.	Brazil	v. *Plants containing photosensitive compounds*: Among the well-studied photosensitive compounds are furocoumarins present in (1) *Psoralea corylifolia*, (2) *Conilla glauca*, and (3) *Ficus carica.* Another plant, (4) *Hypericum perforatum*, also exhibits such properties because of the presence of Hypericin and its analogues. Plant 4 (called Cha-de-Figo locally, used commercially as a tanner and in traditional medicine for the treatment of vitiligo) has been reported to be responsible for several toxic accidents and deaths in Brazil. vi. *Miscellaneous*: Toxic accidents have been reported from some plants, for example: (1) *Diphenbachia* spp., a decorative plant that caused poisoning in children due to the presence of calcium oxalate raphides; (2) *Ruta graveolens* (Arruda) causes abortion; (3) *Atropa* spp.; (4) *Brugmansia*; (5) *Datura* spp. containing toxic alkaloids: Atropine, Scopolamine, and so on, caused atropinic hallucinations; (6) *Chenopodium ambrosides*; (7) *Artemisia absinthium*; (8) *Equisetum* spp. and so on.		
7.	Canada	11. Pahwa and Chatterjee from the University of Saskatchewan, Saskatoon, studied the toxic effects of *Thevetia neriifolia* Juss. (Yellow Oleander) in the roof rat. Oral intake of crushed ground kernels (20% and 30% mixed in feed, for 30 days) exhibited: a. *Neurotoxic effects*: Hind limb paralysis, rolling of body on long axis, circular falling of the tail, muscular twitch, tremors, tetanic convulsions, collapse, and death (in 16 out of 20 and 18 out of 20 rats at two dose levels used). b. *Weight loss*: Significant reduction in weight of surviving rats. c. *Hematological changes*: Reduction in hemoglobin, red blood cells (RBC), total leucocyte count (TLC), and neutrophils and increase in lymphocytes.	Not identified (Crushed grounded seed kernels used)	Pahwa and Chatterjee 1990

(Continued)

TABLE 3.1 (CONTINUED)
Global Picture: 59 Examples of the Adverse Effects of Single-Component and Multicomponent Herbal Drugs from 27 Countries

S. No.	Country	Toxic and Adverse Side Effects or Interactions	Toxic or Active Principle	References
		d. *Biochemical or enzymatic changes*: Significant reduction in blood glucose and serum proteins and increase in blood urea nitrogen (BUN), serum glutamic oxaloacetic transaminase (SGOT), and LDH levels.		
		e. *Histopathological changes*: Inflammatory and degenerative changes in liver included moderate fatty changes, congestion, hepatocytosis, nuclear degeneration, pyknosis, and necrosis in the liver. Proliferation of glomerular endothelium, hypercellularity of the glomerulus, necrosis of convoluted tubular epithelium, anucleosis, and pyknosis in the kidneys. Atrophic erosions and inflammatory changes were also observed in the gastric mucosa.		
		12. *Phytomedicinal informatics approach*: Toxicity of phytochemicals, plant-based extracts, dietary supplements, and medicinal plants in general is of medical importance and must be considered in other plant uses. Chen and Viera from Simon Frazer University, Burnby, described a phytomedicinal informatics approach introducing a novel rAi^2 parameter that incorporates information from large databases to identify and quantify the relative toxic effects (and other effects for comparison) of a group of medicinal plants and the relative toxicity of a given plant. The aim is to provide a clinician or a researcher seeking quantitative indication of the best-established or most-researched phytotoxic effect of the plant of his interest.		Chen and Viera 2010
		a. *Plants*: Seven medicinal plants were selected for the study: (1) *Allium sativum*, (2) *Echinacea purpurea*, (3) *Momordica charantia*, (4) *Aristolochia fangchi*, (5) *Larrea tridentata*, (6) *Zingiber officinale*, and (7) *Vaccinium myrtillus*. These were analyzed for nephrotoxicity and hepatotoxicity. To better illustrate the method and provide a larger base for comparison, other parameters related to chronic medical conditions were also included, for example, nausea, diabetes, and inflammation.		

(*Continued*)

TABLE 3.1 (CONTINUED)
Global Picture: 59 Examples of the Adverse Effects of Single-Component and Multicomponent Herbal Drugs from 27 Countries

S. No.	Country	Toxic and Adverse Side Effects or Interactions	Toxic or Active Principle	References
7.	Canada	b. *Results*: The examples provided by the study identified two plants with major hepato- and nephrotoxic effects and quantified the relative evidence for all of the plants that were analyzed. These plants are *Aristolochia fangchi* for nephrotoxicity and *Larrea tridentata* for hepatotoxicity. Because these databases are dynamic, the author emphasized the importance of a need to re-evaluate and associate parameters (e.g., rAi2) on a regular basis to incorporate the results of novel studies.		
		13. Recently, Luyckx from the Division of Nephrology and Immunology, University of Alberta, presented an extensive review on nephrotoxicity due to alternative medicines. A summary of his findings and comments is given below: a. *Widespread usage*: The use of alternative medicine is highly prevalent around the world. In many areas, this is so because of a lack of access or trust in Western medicine and because these medicines, being "natural," are considered safe. b. *Nephrotoxicity*: The kidney is particularly vulnerable to the toxic insult by virtue of its anatomy and function. Reports of nephrotoxicity associated with alternative medicines encompass all forms of renal dysfunction ranging from electrolyte abnormalities and proteinuria to acute and chronic kidney damage, renal failure, and death. The researcher tabulated 56 alternative remedies (43 of herbal origin and 13 mineral or animal origin drugs and formulations from 35 countries covering all continents) to list their indications, form of acute kidney injury, toxic compounds, adverse effects on other organs or functions, and outcome following medicinal intervention.	Aristolochic acid	Luyckx 2012

(Continued)

TABLE 3.1 (CONTINUED)
Global Picture: 59 Examples of the Adverse Effects of Single-Component and Multicomponent Herbal Drugs from 27 Countries

S. No.	Country	Toxic and Adverse Side Effects or Interactions	Toxic or Active Principle	References
		A prominent example cited was about (1) aristolochic acid nephropathy, which developed into (2) a small outbreak of chronic kidney disease (CKD) in Belgium, (3) then as the pathophysiological mechanism for previously unexplained endemic renal disease, and now (4) as a significant risk factor for CKD in population-based studies. It illustrates the success of diligent collaboration and traditional research. c. *Comments and conclusion*: Mechanisms of alternative medicine-associated kidney disease include direct nephrotoxicity, which may be augmented by underlying predisposing conditions such as dehydration, contamination, adulteration, inappropriate use, and interactions with other medicines.		
8.	China	14. Chronic administration of *Zingiber officinale* Roscoe (ginger) powder (0.5–2.0 g kg^{-1} daily by gastric lavage for 35 days) revealed (1) no mortality, abnormalities in general condition, growth, behavior, food and water consumption, hematological picture, blood biochemical parameters (except a decrease in serum LDH activity in male rats), and overt organ abnormalities. (2) At a high dose (2 g kg^{-1}), there is a significant reduction in the absolute (14.4%) and relative (14.5%) weight of testes. The study by Rong et al. from Guangzhou University of Chinese Medicine provides an insight into the toxicological properties of ginger following long-term exposure.	Not identified (crude root powder used)	Rong et al. 2009

(*Continued*)

TABLE 3.1 (CONTINUED)
Global Picture: 59 Examples of the Adverse Effects of Single-Component and Multicomponent Herbal Drugs from 27 Countries

S. No.	Country	Toxic and Adverse Side Effects or Interactions	Toxic or Active Principle	References
8.	China	15. A recent study reported cytokines as potential biomarkers for liver toxicity induced by *Dioscorea bulbifera* L., a traditionally used herbal medicine in China. Mice given ethyl acetate extract (EF) of the plant 400 mg/kg for 12 days showed marked elevation in serum alanine or aspartate transaminase levels. Histopathological assessment further confirmed liver injury in EF-treated mice. The authors carried out investigations using cytokine-antibody assays, enzyme-linked immunoassays, and demonstrated altered expression of CD30 ligand and interleukin-3 as potential biomarkers for hepatotoxicity that is induced by *D. bulbifera*.		Sheng et al. 2014
9.	France	16. Mostera-Kara et al. from the Hepatogastroenterology Service and Regional Drug Surveillance Centre, St. Antome Hospital, Paris, reported a case of fatal hepatitis with an herbal tea containing *Teucrinum chamaedrys* (Wild Germander) and *Camellia rhea*; the poisoning was attributed to the former. A 68-year-old woman was presented with a four-week history of jaundice and fatigue. Laboratory investigations revealed high bilirubin (311 μmol L^{-1}) alanine transaminase (ALT 900 μ L^{-1}; normal value ≤ 30). She had been taking Tealine (content mentioned above) along with dexfenfluramine three tablets (45 mg) daily for 7–14 days, six weeks before hospitalization. It was the second administration of Tealine (at the same dosage for about two weeks) six months earlier. Tests for tissue antibodies (antinuclear, smooth muscle, antireticulum, and antimitochondrial) were all negative. Ultrasonography revealed no hepatic or biliary abnormalities at that time. Over the next two weeks, the condition worsened (Prothrombin and Factor < 10%), and she became comatose. Liver transplantation was not done because of her age, and she died. Massive hepatic necrosis was detected	Not identified (herbal tea used)	Mostera-Kara et al. 1992

(Continued)

TABLE 3.1 (CONTINUED)
Global Picture: 59 Examples of the Adverse Effects of Single-Component and Multicomponent Herbal Drugs from 27 Countries

S. No.	Country	Toxic and Adverse Side Effects or Interactions	Toxic or Active Principle	References
		following necropsy. The authors cited references to show (1) no reports of hepatotoxicity with dexfenfluramine and (2) *C. rhea* and for (3) well-recognized hepatotoxic effects of Wild Germander. In April 1992, the French Department of Health prohibited the sale of this herbal drug after 26 similar cases of acute hepatitis were reported.		
10.	Germany	17. Renter reviewed the pharmacology and medical applications of *A. sativum* and *A. ursinum* covering all aspects of their pharmaceutical preparations and isolated compounds. The review embodied a detailed description of their pharmacokinetics, pharmacodynamics, and metabolic effects (lipid metabolism, antioxidative, free radical-scavenging, cardiovascular, hematological, antibacterial, antifungal, antiviral, hypoglycemic, antiplatelet-aggregation or adhesive fibrinolytic, antimutagenic, and antitumor effects), medicinal uses, toxicity or side effects, and clinical studies.		Renter 1995
		18. Coumarin is a secondary phytochemical with hepatotoxic and carcinogenic properties. The possibility of a genotoxic mechanism in the carcinogenic effect of coumarin was discounted by the European Food Safety Authority in 2004. It was based on new evidence and allowed tolerable daily intake (TDI) for the first time. A value of 0.1 mg kg^{-1} was arrived on the basis of hepatotoxicity data in animals. Abraham et al. from the Federal Institute of Risk Assessment reviewed the available information with a focus on clinical data in human beings. Hepatotoxicity data from patients, treated with coumarin as a medicinal drug, are available. It revealed a subgroup of the human population that is more susceptible to hepatotoxic effects than the animal species that were investigated.	Coumarins	Abraham et al. 2010

(Continued)

TABLE 3.1 (CONTINUED)

Global Picture: 59 Examples of the Adverse Effects of Single-Component and Multicomponent Herbal Drugs from 27 Countries

S. No.	Country	Toxic and Adverse Side Effects or Interactions	Toxic or Active Principle	References
10.	Germany	The cause for such susceptibility is, at present, unknown. The TDI value was arrived at from extrapolation of animal data to man, using an intraspecies factor of 10 and interspecies factor of 2.5 (with no kinetic subfactor), may be misleading in risk assessment. The high susceptibility of relevant subgroup may be due to dynamic causes that are not covered by the above criteria. The authors stated that the comparison of toxicological data from animals and humans demonstrates that the toxicological mechanisms involved may not be identical. For example, in contrast to humans, the CYP2A6-mediated detoxification of 7-hydroxycoumarin is only a minor pathway in many animal species including rodents and dogs. Nutritional exposure to coumarins is mainly due to consumption of cinnamon (*Cassia cinnamon*) and foods containing cinnamon (tea, cereals, cookies, almond cookies, desserts, chocolates, and wine). Such consumption increases considerably during the winter and Christmas season when markets are flooded with cinnamon-rich food items because of its typical taste and flavor. The coumarin content (mean mg kg^{-1}) is as follows: Cinnamon (2680), tea with cinnamon (231.3), star cookies (37.7), cereals (25.5), almond cookies (16.2), gingerbread cake (10.3), dessert with cinnamon (10.2), chocolate with cinnamon (9.4), and malted wine (0.2).		

(Continued)

TABLE 3.1 (CONTINUED)
Global Picture: 59 Examples of the Adverse Effects of Single-Component and Multicomponent Herbal Drugs from 27 Countries

S. No.	Country	Toxic and Adverse Side Effects or Interactions	Toxic or Active Principle	References
10A.	Germany, Italy, and Japan (collaborative study)	19. A recent collaborative study by researchers from Frederich Alexander University, Eriangen (Germany), Centro Cardiologico Monzino, Milan (Italy), Fukushima Medical University, Fukushima (Japan), and Hamamatsu University, Hammamatsu (Japan), reported the marked interaction of green tea (GT) with antihypertensive drugs. An evaluation of pharmacokinetics and pharmacodynamics of the β-blocker drug Nadolol revealed that concentrations of Nadolol were greatly reduced following ingestion of GT in healthy subjects. Ten healthy volunteers received a single oral dose (Nadolol 30 mg) with GT or water. After repeated consumption of GT (700 mL per day) or water for 14 days, GT markedly decreased Cmax and AUC_{0-48} of Nadolol by 85.3% and 85%, respectively ($p < 0.01$), without altering its renal clearance. Further, the effects of Nadolol on blood pressure were significantly reduced by GT. The effects were possibly elicited in part by inhibition organic anion-transporting polypeptide-mediated uptake of Nadolol in the intestine.	–	Misaka et al. 2014
11.	Ghana	20. *Aconitum* spp. (*A. carmichaeli* and *A. kusnezoffi*) are traditionally used for relief of pain (China Pharmacopoeia Committee 2010). Diester diterpenes, contained in these plants, are responsible for their toxic effects. Severe cases of poisoning (cardiotoxicity) due to homemade Chinese remedies containing *A. carmichaeli* have been reported. Cardiotoxic effects include tachycardia, bradycardia, hypotension, arrhythmias, fibrillation, and death.	Aconitine, mesaconitine, and Hypaconitine	Liu et al. 2011; Ekor 2013

(*Continued*)

TABLE 3.1 (CONTINUED)
Global Picture: 59 Examples of the Adverse Effects of Single-Component and Multicomponent Herbal Drugs from 27 Countries

S. No.	Country	Toxic and Adverse Side Effects or Interactions	Toxic or Active Principle	References
11.	Ghana	21. *Allium sativum* (Garlic), used as food or as a food additive, has also found relevance for the management of hypertension and hypercholesterolemia. It is known to contain Allin, which upon crushing or chopping without heat or acid gets activated by Allinasa and becomes Allicin. Adverse effects associated with garlic extract include a burning sensation in the gastrointestinal tract (GIT), nausea, diaphoresis, light-headedness, and contact dermatitis. Excessive ingestion of garlic has been reported to cause morbid spontaneous spinal epidural hematoma.	Allin, Allicin	Rose et al. 1990; Ekor 2013
		22. *Ephedra sinica* (Ephedra) is a popular herb with a long history of traditional use in respiratory ailments. A number of randomized double-blind clinical trial reports are available on this herb. It finds a place in the *Chinese Pharmacopoeia* (2010). Many adverse reports have linked *E. sinica* to hepatotoxicity, neurotoxicity, and blindness.		Ekor 2013
		23. *Cryptolepis sanguinolenta* root decoction is widely used for the treatment of malaria. Clinical efficacy of its teabag formulation in the treatment of acute uncomplicated *falciparum* malaria has been reviewed. However, it has recently been shown to have cytotoxic and other metabolic properties.	Cryptolepine (major alkaloid)	Bugyei et al. 2010; Ansah and Mensah 2013

(Continued)

TABLE 3.1 (CONTINUED)
Global Picture: 59 Examples of the Adverse Effects of Single-Component and Multicomponent Herbal Drugs from 27 Countries

S. No.	Country	Toxic and Adverse Side Effects or Interactions	Toxic or Active Principle	References
12.	Hong Kong	24. But from the Chinese University of Hong Kong stressed the need for correct identification of herbs in herbal drug poisoning cases. He cited several references to support this contention. The commonly used names are very vague and confusing, for example, Ginseng, Ren Shen, Radix Ginseng, and so on. The name Ginseng is applied to *Panax ginseng* (Oriental Ginseng), *Panax quinquefolia* (American Ginseng), and *Eleutherococcus senticosus* (Siberian Ginseng). Scientific names are more specific and acceptable versus common names, transliterations, and pharmaceutical names. The situation is further complicated by the fact that many herbs imported to or re-exported from Hong Kong may be derived from sources other than those in the referred work. Fang-Ji, re-exported from Hong Kong, is largely derived from *Aristolochia fanchi* (not from *Stephania tetrandra* or *Coculus trilobus*). *A. fangchi* has caused nephrotoxic reaction. The Chinese Medicinal Material Research Center can be contacted for authentication of such drugs.		But 1993
		25. A multidisciplinary approach to the toxicological problems associated with the use of herbal medicines was introduced in Hong Kong in 2000. A team consisting of a pharmacist, a chemical pathologist, a scientific officer, and a physician jointly provides advisory service on herbal safety to healthcare professionals. The first two-year experience of this team was reviewed by researchers from the Drugs and Poisons Information Bureau, Chinese University of Hong Kong, and Prince of Wales Hospital. It discussed several aspects for 25 referrals that were received from hospitals during a two-year period: precise diagnosis for the temporal link between exposure and adverse event, exclusion of other causes, pharmacological or toxicological and analytical studies to distinguish between causal relationship and conclusive link, clinical problems and their mitigation or control, etc., *Aristolochia* spp., *Aconite datura* spp., and *Stephania sinica* (Yulan) were described for renal failure, aconitine poisoning, anticholinergic effects, and tetrahydropalmatine poisoning, respectively.		Chan et al. 2005; Li et al. 2012; Wang et al. 2012

(Continued)

TABLE 3.1 (CONTINUED)
Global Picture: 59 Examples of the Adverse Effects of Single-Component and Multicomponent Herbal Drugs from 27 Countries

S. No.	Country	Toxic and Adverse Side Effects or Interactions	Toxic or Active Principle	References
12.	Hong Kong	Safety evaluation of commonly used medicines during pregnancy in mice and adverse outcomes of Chinese medicines used for threatened miscarriage were reviewed recently. Meta-analysis for the latter revealed a considerable variation in design and interventions of outcome measures; therefore, the results remain elusive. The reviewers stressed the need for rigorous scientific and placebo-controlled trials for safety of the masses.		
13.	Iran	26. Talasaz et al. from Tehran University of Medical Sciences reported a case of *Tribulus terrestris*-induced severe nephrotoxicity–hepatotoxicity and neurotoxicity in a young Iranian male, who used the plant extract to prevent kidney stone formulation. He presented with seizure symptoms and very high serum aminotransferase levels and creatinine following consumption of herbal water for two days. Discontinuation of the herbal remedy resulted in improvement of symptoms and normalization of liver enzymes.		Talasaz et al. 2010
		27. *Xanthium strumarium* (Cocklebur): The mature plant is considered nontoxic, but burrs are toxic to domestic animals (cattle, horses, pigs). Intake of plants with burrs may be harmful—not limited to mechanical injury during mastication. Glycoside poisoning causes multiple organ dysfunction, coagulation abnormalities, low blood glucose and potassium levels, hepatic and renal failure, and death in livestock. There is no antidote for it, and supportive measures are the mainstay of treatment. Some cases of human poisoning have also been reported. Patients with acute onset show severe abdominal pain, nausea, vomiting, drowsiness, sweating, palpitation, and dyspnea. These may progress to loss of consciousness and death. Allergenic compounds in pollens may cause contact dermatitis in atopic patients.	a. Carboxyatractyloside: glycoside present in seeds b. Allergenic compounds in pollens	Saidi and Mofidi 2009 Faghihi and Radan 2011

(Continued)

TABLE 3.1 (CONTINUED)
Global Picture: 59 Examples of the Adverse Effects of Single-Component and Multicomponent Herbal Drugs from 27 Countries

S. No.	Country	Toxic and Adverse Side Effects or Interactions	Toxic or Active Principle	References
		Saidi and Mofidi from the Department of Emergency Medicine, Hazrat Rasool Hospital, Tehran, described a case of intoxication resulting from exposure to this plant in a previously healthy young (25-year-old) woman who was taking a herbal preparation from this plant. The woman was referred to the department with altered mental status and an episode of tonic and clonic seizures. Intoxication was reported within three days of drinking a decoction of the plant, and convulsions developed on day 7. Laboratory investigations, sonography, and CT scan revealed rapid hypoglycemia (improved on intravenous dextrose administration), prolonged coagulation profiles, elevated hepatic and renal enzymes, hepatitis, edema, and ascites. Hospitalization for seven days and subsequent monitoring for three months led to complete recovery. She did not take this herbal decoction afterward.		
		28. Faghihi and Radan from the Department of Dermatology, University of Medical Sciences, Isfahan, reviewed 25 commonly used herbal drugs for dermatological problems and in cosmetic preparations. Because of poor monitoring, adverse reactions (mild to severe) may occur following exposure to these drugs when used alone or in combination with other herbal or modern drugs. The symptoms include contact dermatitis, urticaria, allergic reactions, rashes, irritation, itching, phototoxicity, coagulation abnormalities or bleeding, hepatotoxic, central nervous system (CNS) problems, anaphylaxis, hormonal side effects, and so on. Some examples are listed below: a. *Aloe vera* gel (juice) should not be used in cases with deep vertical cuts as it may delay healing and cause allergic reactions in some people. b. Fenugreek causes hypoglycemia. c. Licorice should not be used in patients with hypertension. It induces hypokalemia and an aldosterone-like effect.	a. Psoralens in celery b. Hepatotoxic alkaloids in Comfrey	

(Continued)

TABLE 3.1 (CONTINUED)
Global Picture: 59 Examples of the Adverse Effects of Single-Component and Multicomponent Herbal Drugs from 27 Countries

S. No.	Country	Toxic and Adverse Side Effects or Interactions	Toxic or Active Principle	References
13.	Iran	d. Ephedra (incorporated in some weight loss products) is like amphetamine and may cause CNS problems. e. *Panax ginseng* should not be used with stimulants, antipsychotic drugs, and in patients with diabetes. f. Celery and green tea may interact with warfarin and increase the risk of bleeding. g. Celery contains psoralens that may cause a toxic dermal reaction on exposure to ultraviolet A rays. h. Henna, considered harmless, may cause severe allergic reactions in some individuals. Even cases of systemic toxicity or anaphylaxis have been reported with this herb. i. Sage may cause tenderness in breasts and irregular menses. j. Accidental oral ingestion of herbal drugs for topical use may cause serious toxic effects. It is important for dermatologists to become aware of these adverse effects and interactions in order to educate their patients regarding the risks that are involved and possibly prevent potential and unexpected herbal drug toxicity. Caution is especially needed in pregnant women, infants, and in patients with other diseases.		
14.	Italy	29. Grieco et al. from the Institute of Internal Medicine, Catholic University of Rome, reported a case of acute hepatitis that was caused by a natural lipid-lowering product Equisterol. A 63-year-old woman presented with severe hypertransaminasemia that developed progressively over a few weeks. She had been taking this over-the-counter herbal product for six months. It contains guggulsterol (from *Commiphora mukul*) and red yeast rice extract. The product had been prescribed to treat hypercholesterolemia because the patient had developed hepatotoxicity while on Lovastatin.	Guggulsterol	Grieco et al. 2009

(Continued)

TABLE 3.1 (CONTINUED)
Global Picture: 59 Examples of the Adverse Effects of Single-Component and Multicomponent Herbal Drugs from 27 Countries

S. No.	Country	Toxic and Adverse Side Effects or Interactions	Toxic or Active Principle	References
		A liver biopsy revealed severe lobular necroinflammatory changes with eosinophilic infiltration. The episode was regarded as an adverse drug reaction after (1) exclusion of other possible causes of acute liver disease and (2) prompt normalization of liver function tests following discontinuation of Equisterol. Red yeast rice extract's cholesterol-lowering properties are largely due to monocolins (one of which is Monocolin-k identical to Lovastatin). The researchers question the choice of alternative medicine to "rechallenge" with official medicinal agents that had previously caused mild hepatotoxicity. They concluded that the physician should keep in mind that alternative medicine is not always the safest alternative, and sometimes, it is not even an alternative.		
		30. Cravotto et al. from the University of Jurin, in an extensive review, evaluated the pharmacotherapeutic potential of 1000 plants using bibliographic sources from several databases (updated to October 2007) with the aim to summarize the available scientific information on commonly marketed plant extracts in Western countries. They paid particular attention to new plants with powerful anti-inflammatory activity. In order to assess the value of clinical trials, they focused on nine plants with the maximum trial data available: (1) *Althea officinalis*, (2) *Calendula officinalis*, (3) *Centella asiatica*, (4) *Echinacea purpura*, (5) *Passiflora incarnate*, (6) *Punica granatum*, (7) *Vaccinium macrocarpum*, (8) *Vaccinium myrtillus*, and (9) *Valeriana officinalis*. These plants had considerable evidence for therapeutic effect. The survey revealed paucity of clinical data; clinical trial reports were published only for 156 out of 1000 plants. However, for about 50% of the plants surveyed, *in vitro* and *in vivo* studies provided some support for their therapeutic use. There was strong evidence that 1 in 200 plants was toxic or allergenic.		Cravotto et al. 2010

(*Continued*)

TABLE 3.1 (CONTINUED)
Global Picture: 59 Examples of the Adverse Effects of Single-Component and Multicomponent Herbal Drugs from 27 Countries

S. No.	Country	Toxic and Adverse Side Effects or Interactions	Toxic or Active Principle	References
14.	Italy	Their use should be discouraged or forbidden. Establishing the efficacy and safety of most phytochemicals and identifying their mechanism(s) of action and toxicity are major challenges. The reviewers provided a baseline on the level of evidence available to help intending researchers who are working on these topics.		
		31. Rhabdomyolysis is an uncommon side effect of Trabecedin (TRB), which is used as a second-line therapy of metastatic sarcoma after Anthrocycline and Ifosfamide failure. This side effect may be due to pharmacokinetic interactions that are caused by shared metabolism involving cytochrome P450 system in the liver. A 56-year-old Caucasian affected by a relapsed dedifferentiated liposarcoma complained of sudden weakness, difficulty in walking, and diffuse muscular pain. Upon hospital admission, the patient showed grade (G) 4 pancytopenia and a marked increase in liver lytic enzymes, serum myoglobin, creatinine phosphokinase (CPK), and LDH. No cardiac or kidney function injuries were present. The patient was diagnosed with rhabdomyolysis induced by TRB. He had not received any drugs during treatment with TRB, but upon further questioning, the patient revealed that he had been taking a folk medicine preparation of Chokeberry (*Aronia melanocarpa*) daily during the last course of TRB and in the two subsequent weeks. One week after hospitalization and cessation of Chokeberry intake, CPK and other markers of myolysis slowly returned to standard range. The patient noted progressive recovery of muscle strength and mobility. The patient was discharged on day 14 when a blood transfusion and parenteral hydration gradually lowered the general toxicity that was confirmed by laboratory findings.		Strippoli et al. 2013

(*Continued*)

TABLE 3.1 (CONTINUED)
Global Picture: 59 Examples of the Adverse Effects of Single-Component and Multicomponent Herbal Drugs from 27 Countries

S. No.	Country	Toxic and Adverse Side Effects or Interactions	Toxic or Active Principle	References
		Researchers from the National Cancer Research Center, Giovanni Paolo II, Brazil, claimed it to be a first report of herbal drug interaction rhabdomyolysis with TRB. It underlines the importance of understanding rare treatment-related toxicities and interactions.		
15.	Japan	32. Several reports have focused on adverse reactions or poisoning due to traditional Chinese herbal remedies (TCHRs). In Japan, TCHRs, originally imported from China, have evolved independently and are referred to as Kempo medicines. Over 100 types of drug extracts are available commercially in Japan, and Kempo remedies are accepted by the Japanese National Health Insurance System. The quality of herbs is regulated by determining the main chemical constituents according to guidelines in the *Japanese Pharmacopoeia*. Most Japanese doctors and patients feel comfortable in using them even when these drugs are taken at the same time with Western medicine. Homma et al. from the Department of Pharmacology, College of Pharmacy, Tokyo, stated that the efficacy and toxicity of Kempo medicines are uncertain because double-blind. placebo-controlled trials have been rare. They attributed this to difficulties in preparing *placebos* with the unique smell and taste of such herbal drugs or formulations. In order to find out which chemicals are absorbed into the body from Kempo medicines, these researchers investigated Saiboku-to that was used in cases of asthma for steroid-sparing effect: it contains 10 herbal extracts; 7 chemicals were isolated from the urine of subjects after ingestion of Saiboku-to: a. Magnalol from *Magnolia officinalis*, b. Liquiritigenin and Isoliquiritigenin from *Glycyrrhiza glabra*, c. Wagonin, Baicolenin, and Orxylin-A from *Scutelleria baicalensis*, and d. Bavidigenin of unknown origin.	Magnolol	Homma et al. 1993

(Continued)

TABLE 3.1 (CONTINUED)
Global Picture: 59 Examples of the Adverse Effects of Single-Component and Multicomponent Herbal Drugs from 27 Countries

S. No.	Country	Toxic and Adverse Side Effects or Interactions	Toxic or Active Principle	References
15.	Japan	Surprisingly, major chemicals, found in the plants, could not be detected in urine or blood: the chemicals isolated from urine are lignans and flavonoids; Magnolol was unique because the levels of conjugated form were notably higher in responders to Saiboku-to treatment versus those in the nonresponders who were suggesting this chemical to be responsible for its steroid-sparing activity. These workers examined the pharmacodynamics of Magnolol in relation to steroid metabolic enzyme inhibition. It inhibited 11β-hydroxysteroid dehydrogenase, a key enzyme for glucocorticoid metabolism. Though its activity was less than that of Glycyrrhizin (a major component of *G. glabra*), it was hardly absorbed. The authors emphasized that pharmacokinetic investigations are the key to resolving the mysteries of Kempo medicines.		
16.	Malaysia	33. In a continuing pharmacy education (CPE) program, Hussin from the School of Pharmaceutical Sciences, Universiti Sains Malaysia, discussed various aspects of adverse effects and drug–herbal interactions including widespread global usage, increasing sales, reasons for people resorting to herbal therapy, factors affecting toxicity, and risk groups followed by questions for CPE. a. *Plants*: (1) *A. sativum* (Garlic), (2) *Datura fastuosa* (Kekubong), (3) *Datura metel* (Kekubong), (4) *Datura stramonium* (Terang Pengas), (5) *Eucalyptus globus* (Eucalyptus), (6) *Ganderma myelinum* (Kalat Kayis), (7) *Ginkgo biloba*, (8) *Hypericum perforatum* (St. John's Wort), (9) *Larreatri dentate* (Chapparal), (10) *Melia azadirachta* (Margosa or Darin Mamba), (11) *Mentha pulegium* (Pennyroyal), (12) *Marinda citrifolia* (Noni or Menga Kadu), (13) *Myristica fragrans* (Nutmeg), (14) *Parica speciose* (Petai), (15) *Piper methysticum* (Kava Kava), (16) *Pithocallobium jiringa* (Jering), (17) *Pithocolobium microcarpus* (Kerdas), (18) *Symphytum* spp. (Comfrey), (19) *Tecurium chamaedrys* (Germander), (20) *Valleriana officinalis* (Valerian), and (21) *Zingiber officinale* (Ginger).	a. Pyrrozolidine alkaloids b. Steroids in 25% unregistered preparations	Hussin 2001

(*Continued*)

TABLE 3.1 (CONTINUED)
Global Picture: 59 Examples of the Adverse Effects of Single-Component and Multicomponent Herbal Drugs from 27 Countries

S. No.	Country	Toxic and Adverse Side Effects or Interactions	Toxic or Active Principle	References
		b. *Herbals containing toxic ingredients*: i. Pyrrozolidine alkaloids in plants 9, 18, and 19 were found to be associated with hepatotoxicity in many cases. ii. The oil in plant 11 contained an abortifacient compound with reported hepatotoxic and renal toxic effects, which may lead to death. iii. Plant 12 with high-potassium content should be used with caution in patients with renal problems. It may cause arrhythmias and myocardial infarction in many cases. c. *Unintentional or intentional substitution (adulteration) with other drugs*: Many of the locally available herbal preparations contained the wrong medicinal plant due to misidentification or substitution with cheaper alternatives or easily available species. Adulteration with steroids was found in the majority of herbal products in Malaysia. The authors cited an example from China. d. *Toxic interactions with modern medicine*: i. Interactions with prescribed anticoagulants or antiplatelet drugs (e.g., aspirin, coumarin, heparin, warfarin) were observed in plants 1, 7, and 21 exposing the patient to the risk of bleeding. A case report showed spontaneous bleeding from the iris within a week of *G. biloba* supplementation in a patient who had been taking aspirin to prevent cardiac disease. Unexpected bleeding has been observed in patients during surgery due to consumption of such herbal products. ii. Plants 8, 15, and 20 may prolong the effect of some anesthetics and delay awakening (potentiation). iii. Some other drugs may decrease (antagonism) the intended effect of prescribed medicines. Plant 8 caused a decrease in the efficacy of drugs that are used to treat AIDS, cancer, and birth control (about 50% decrease was noted in the effect of birth control pills).		

(*Continued*)

TABLE 3.1 (CONTINUED)
Global Picture: 59 Examples of the Adverse Effects of Single-Component and Multicomponent Herbal Drugs from 27 Countries

S. No.	Country	Toxic and Adverse Side Effects or Interactions	Toxic or Active Principle	References
16.	Malaysia	iv. Concurrent use of herbal products with modern medicine may alter their absorption, distribution, metabolism and excretion, pharmacokinetic, or pharmacodynamic properties with serious consequences. The researchers reported that plants 14 and 17 caused a significant alteration in hepatic drug metabolism *in vitro* and *in vivo*.		
17.	Mauritius	34. A recent review presented the toxicological challenges associated with the use of herbal medicinal products in developing countries. She addressed various aspects of the problem. These included natural health product regulations, protocols, and guidance documents on safety and toxicity testing. Neergheen-Bhujan from the University of Mauritius, Reduit, Mauritius, listed several examples of (1) selected plants traditionally used as single herbs and (2) evidence-based herbal drug toxicity and interactions from these countries. Some prominent examples from the latter category are as follows: a. Use of *Sauropus androgynus* in Taiwan and Japan for weight control was associated with an outbreak of obstructive lung disease (Lai et al. 1996). b. Hyperkalemia due to toxic levels of cardiac glycosides was linked to a long list of herbal medicines that are taken by patients (Haden et al. 2011). c. Ten cases of severe hepatotoxicity (including a case of hepatitis E) were associated with dietary supplements from herbal products (Schoepfer et al. 2007). d. *Glycyrrhiza glabra* (Licorice), commonly used for inflammation of the upper respiratory tract and gastric and duodenal ulcers, may cause suppression of the renin–aldosterone system leading to sodium and water retention. This may result in hypokalemia, hypertension, cardiac arrhythmias, and encephalopathy after prolonged use (Russo et al. 2000; Padiyara and Khan 2006). e. The sudden death of a healthy college student was linked to Ephedrine from a Ma Huang-containing drink (Theoharides 1997).		Neergheen-Bhujan 2013

(Continued)

TABLE 3.1 (CONTINUED)
Global Picture: 59 Examples of the Adverse Effects of Single-Component and Multicomponent Herbal Drugs from 27 Countries

S. No.	Country	Toxic and Adverse Side Effects or Interactions	Toxic or Active Principle	References
18.	Morocco	35. Bnouham et al. from the University of Mohammad, Onida, and Rabat Institute reviewed the toxic effects of 28 medicinal plants that are commonly used in Moroccan traditional medicine. a. *Plants*: (1) *Aconitum vulparia* (Igntar, Qatel-ed-Dib), (2) *Adonis oestivalis* (Ayn-al-Hajla, Benoman Saghir), (3) *Angyris foetida* (Rharoub, Ikhenzis, Ful-i-kleb, Tizzals), (4) *Aristolchia longa* (Aarfi, Chajrat Rustam) (5) *Atractylis gummifera* (Addad, Ahfyun), (6) *Atropa belladonna* (Tidilla, Zbib Elkhidus), (7) *Bryonia dioica* (Fabir, Luwaya), (8) *Chenopodium album* (Blis, Remren), (9) *Citrulus colocynthis* (Hamlal, Lehdej), (10) *Colchicum autumnale* (Bukbuka, Tirkit), (11) *Conicumm aculatum* (Sikram, Barbus), (12) *Daphine gnidum* (Lezzar, Metnan), (13) *Daphnel aureola* (Ad-dufayla, Talidras), (14) *Datura stramonium* (Chedqejjmel, Tabrzig), (15) *Digitalis lanata*, (16) *Ferula communis* (Ikelcha, Ufful), (17) *Hyocyamus niger* (Sikram, Benjaraj), (18) *Trispseudo carpus* (Udalambar, Busrika), (*19) *Mandsagara autamnalis* (Bid elghoul, Taryala), (20) *Nerium oleandra* (Odefla, Elel), (21) *Ricinus communis* (Kherwa, Krank), (22) *Solanum nigrum* (Inebeddib, Buquira), (23) *Solanum bodomaeum* (Lim masra, Hedja), (24) *Tamus communis* (Tamier, Belminmoune), (25) *Taxus baccata* (Dahu, Igen), (26) *Thapsia garganica* (Deryar, Buneffar), (27) *Urginea maritima* (Bsal elkhenzi), and (28) *Withania somnifera* (Habal lakw, Sikram). b. *Findings*: i. While all 28 plants (from 16 families) affect physiological functions, 17 plants were considered more dangerous as these caused irreversible health problems and damage to vital functions (blindness, hepatotoxicity, renal failure, hemiplegia, and death in some cases).		Bnouham et al. 2006

(*Continued*)

TABLE 3.1 (CONTINUED)

Global Picture: 59 Examples of the Adverse Effects of Single-Component and Multicomponent Herbal Drugs from 27 Countries

S. No.	Country	Toxic and Adverse Side Effects or Interactions	Toxic or Active Principle	References
18.	Morocco	A. *Aconitum vulparia*: Diarrhea, shivering, hypertension, tachycardia, vertigo, and respiratory paralysis	Aconitine	
		B. *Anagyris foetida*: Gastrointestinal disturbances, diarrhea, vomiting, hypertension, and tachycardia	Anagarine	
		C. *Atractylis gummifera*: Fatal poison	a. Atractosides b. Carboxyatractylosides c. Paraquine d. Carboxyquine	
		D. *Bryonia dioica*: Cramps, hypothermia, convulsions, and coma	a. Brydiofin b. Cucurbitabcin	
		E. *Chenopodium album*: Nausea, vomiting, hypotension, CNS toxicity, cephalgia, meningitis, vertigo, and paralysis	Oxalic acid	
		F. *Colchicun autumnale*: Diarrhea, vomiting, antimitotic action, and general and respiratory paralysis	Colchicine	
		G. *Coniumm aculatum*: Asphyxia and death	a. Coniline b. Conhydrine c. Methyl coninine	
		H. *Daphne gnidum*: Diarrhea, difficulty in deglutination, shivering, headache, palor, pulmonary dysfunction, dilatation of pupils, convulsions, and death	a. Daphetoxin b. Mezirin c. Vesicant resin d. Tannins	

(Continued)

TABLE 3.1 (CONTINUED)
Global Picture: 59 Examples of the Adverse Effects of Single-Component and Multicomponent Herbal Drugs from 27 Countries

S. No.	Country	Toxic and Adverse Side Effects or Interactions	Toxic or Active Principle	References
		I. *Daphnel aureola*: While leaves are used in medicine, berries are toxic. Dilation of pupils, palor, swelling of lips or mouth, diarrhea, difficulty in deglutination, pulmonary dysfunction, convulsions, and death	No compound identified	
		J. *Ferula communis*: Anorexia, diarrhea, internal and external hemorrhages, and antivitamin K activity	a. Ferulinol b. Hydroxy-ferulinol c. Ferprenine d. Iso-ferprenine	
		K. *Hyocyamus niger*: Dryness of mucus membranes; gastrointestinal, respiratory, cardiac, and neurotoxic effects; tachycardia; depression; dyspnea; coma; convulsions; and death	a. Atropine b. Hyoscyamine c. Scopolamine	
		L. *Mandragora autmnalis*: Fatal poison	a. Atropine b. Hyocyamine c. Scopolamine	
		M. *Nerium oleander*: Gastroenteritis, vomiting, shivering, hypothermia, bradycardia, respiratory problems, convulsions, asphyxia, and cardiac failure	a. Oleandrin b. Neriine or Neraside c. Oleandroside d. Ouabain e. Digitoxin	

(*Continued*)

TABLE 3.1 (CONTINUED)
Global Picture: 59 Examples of the Adverse Effects of Single-Component and Multicomponent Herbal Drugs from 27 Countries

S. No.	Country	Toxic and Adverse Side Effects or Interactions	Toxic or Active Principle	References
18.	Morocco	N. *Ricinus communis*: Fatal poison	Ricin	
		O. *Solanum nigrum*: Vomiting, mental confusion, CNS depression, sedation, hallucinations, tachycardia, coma, and death	a. Solanine b. Solanidine	
		P. *Tamus communis*: Irritant to mucus membranes and respiratory problems	a. Campesterol b. Stigmosterol c. β-sisterol	
		Q. *Thapsia garanica*: Gastroenteritis, salivation, colic, intense diarrhea, nervine disorders, and death ii. The most representative family for toxic plants was Solanaceae (7 species) iii. Alkaloidal compounds were found to be the most frequent toxic principle (9 plants) iv. Overdose; wrong usage; substitution or adulteration with other herbal drugs; concurrent use with other herbal drugs or modern medicines (drug interactions); use in diseased states, pregnant, nursing women, infants, children, and elderly persons (risk persons); and a host of other factors affect safety and warrant caution	Resins	

(Continued)

TABLE 3.1 (CONTINUED)
Global Picture: 59 Examples of the Adverse Effects of Single-Component and Multicomponent Herbal Drugs from 27 Countries

S. No.	Country	Toxic and Adverse Side Effects or Interactions	Toxic or Active Principle	References
19.	Netherlands	36. Boeke et al. from the University of Wageningen carried out a safety evaluation of Neem (*Azadirachta indica*) derived pesticides. They stated that in addition to possible beneficial health effects of this plant (e.g., antidiabetic, antiparasitic, antiulcer, hepatoprotective, and pesticidal actions), its toxic effects have also been reported. The authors presented a review of toxicological data from animal and human studies with oral administration of Neem-based preparations. Non-aqueous extracts appear to be most toxic with an estimated safe dose (ESD) of 0.002 and 12.5 μg kg^{-1} per day. Unprocessed materials (seed oil and aqueous extracts) were less toxic (ESD 0.26 and 0.3 mg kg^{-1} per day). Most pure compounds exhibited relatively low toxicity (ESD for Azadiractin: 15 mg kg^{-1} per day). For all preparations, a reversible effect on reproductive function (in both male and female mammals) was found to be the most important toxic effect following subacute and chronic exposure. From a safety assessment of various Neem-derived pesticides or insecticides by ingestion of residues of contaminated food, it was concluded that such usage should not be discouraged.	Azadiractin	Boeke et al. 2004

(Continued)

TABLE 3.1 (CONTINUED)
Global Picture: 59 Examples of the Adverse Effects of Single-Component and Multicomponent Herbal Drugs from 27 Countries

S. No.	Country	Toxic and Adverse Side Effects or Interactions	Toxic or Active Principle	References
20.	Nigeria	37. Adedapo et al. from the University of Ibadan studied the toxic effects of *A. precatorius* in male rats. Oral administration of aqueous extract of leaves (400, 800, and 1600 mg kg^{-1}) for 18 days exhibited marked toxic effects. a. *Hematological alterations*: Decrease in packed cell volume, hemoglobin concentration, RBC and WBC count, mean corpuscular volume, and mean corpuscular hemoglobin content. b. *Biochemical changes*: Increased levels of total serum protein, albumin, alanine amino transferase, aspartate aminotransferase, alkaline phosphatase, and total bilirubin. c. *Histopathological changes*: Testicular degeneration characterized by a decrease in the number of lining cells of the epithelium and reduction in sperm count with presence of scattered sertoli cells. In view of these toxic findings, the researchers advised to exercise caution in the use of this plant for medicinal purposes.	Not identified (crude extracts used for the study)	Adedapo et al. 2007
		38. The methanolic extract of the leaves of *Aspilla africans* (150–300 mg kg^{-1} orally, once daily for 30 days) exhibited significant toxic effects on the estrus cycle, uterine weight, histological changes in uterine muscle, and endometrial glands in Wistar rats. The effects were found to be dose dependent. Further decrease in pro-estrus, estrus phases, and ovulation; deletion or derangement of endometrial stroma; and thickening of endometrial blood vessels were also noted. The results indicate a negative influence of the plant's extract on the reproductive health of animals and potential toxicity possibly due to hormonal changes.	High concentration of phytoestrogens like saponins and essential oils	Oluyemi et al. 2007

(*Continued*)

TABLE 3.1 (CONTINUED)
Global Picture: 59 Examples of the Adverse Effects of Single-Component and Multicomponent Herbal Drugs from 27 Countries

S. No.	Country	Toxic and Adverse Side Effects or Interactions	Toxic or Active Principle	References
21.	Portugal and Brazil (collaborative study)	39. Medicinal plant research is on the rise universally. The tendency is to consider natural products as nontoxic and with fewer side effects versus the conventional medicine; Petronilho et al. from University of Avira, Portugal, and Federal University of Santo Catarina, Brazil, reviewed *in vitro* and *in vivo* studies on natural products. A case study of the plant Chamomile (*Matricaria recutita*) was described in detail to critically evaluate the published data available for its health benefits and safety concerns. Their observations are summarized as follows. a. *Published information* It is one of the most widely used and researched plants worldwide. The authors reviewed the available *in vitro* and *in vivo* studies (antimicrobial, antioxidant, antimalarial, antimutagenic, antiplatelet, antichemotactic, anti-inflammatory, anticancer, and osteoporosis-preventing activities and interference with drug metabolism) and *in vivo* studies in animals and humans. Several health benefits have been reported for the essential oil, crude extracts, and Sesquiterpene compounds that are present in the plant. b. *Utility and drawbacks* While considerable data exist supporting the claims of its medicinal value in *in vitro* studies and for *in vivo* studies in animal models, information concerning the real clinical benefit for human health is insufficient and inadequate. A deep knowledge concerning the components of Chamomile for each specific effect and the mechanisms involved have not been established in most of the reports. Sesquiterpene compounds seem to be involved in a network of biological effects; they may exhibit synergistic effects with other drugs such as antibiotics and/or may alter the metabolism of certain coadministered drugs and mutagenic effects. The researchers concluded the following:	a. Essential oil b. Sesquiterpene compounds (α-Bisabolol and its oxides, α- and β-Farnesene, Germacrene D, etc.)	Petronilho et al. 2012

(*Continued*)

TABLE 3.1 (CONTINUED)
Global Picture: 59 Examples of the Adverse Effects of Single-Component and Multicomponent Herbal Drugs from 27 Countries

S. No.	Country	Toxic and Adverse Side Effects or Interactions	Toxic or Active Principle	References
21.	Portugal and Brazil (collaborative study)	i. *In vitro* and animal studies data are very promising for the potential health benefits of this plant in varied conditions, and ii. More efforts are needed for controlled or clinical trials in patients to explore the potential.		
22.	South Africa	40. *Illicium verum* (Chinese Star Anise) dried fruit is popularly used as a remedy to treat infant colic. However, instances of life-threatening adverse events have been recorded after use in some cases due to substitution and/or adulteration with *Illicium anisatum* (Japanese star anise), which is toxic. It is evident that rapid and efficient quality control methods are of utmost importance to prevent recurrence of such dire consequences. With this aim in view, researchers from Tshwane University of Technology, Pretoria, developed short-wave infrared as an objective and nondestructive method to distinguish between dried fruits of *I. verum* and *I. anisatum*.		Vermaak et al. 2013
22A.	South Africa and Nigeria (collaborative study)	41. Recently, Ashafa et al. investigated the toxicity profile of the ethanolic extract of stem bark of *Azadirachta indica* in male Wistar rats. The study was carried out in 50, 100, 200, and 300 mg kg^{-1} po for 21 days. While a low dose (50 mg kg^{-1}) appears to be safe, higher doses caused significant alterations in biochemical and other parameters of toxicity including body or organ weight indicating adverse effects on functioning of various organs. It was concluded that the ethanolic extract of this plant may not be a safe oral remedy at doses of 100–300 mg kg^{-1}.	Ethanolic extract	Ashafa et al. 2012

(*Continued*)

TABLE 3.1 (CONTINUED)
Global Picture: 59 Examples of the Adverse Effects of Single-Component and Multicomponent Herbal Drugs from 27 Countries

S. No.	Country	Toxic and Adverse Side Effects or Interactions	Toxic or Active Principle	References
23.	Spain	42. Garcia et al. the from the University of Leon carried out experimental studies to probe the pharmacokinetic interaction between hydrosoluble fiber (*Plantago ovata* husk) and levodopa. While a lower dose of fiber (100 mg kg^{-1}) did not affect the absorption of levodopa (20 mg kg^{-1}) on oral administration, it increased the extent of absorption and mean residence time of levodopa in rabbits. The authors concluded that *P. ovata* husk administration with levodopa could be beneficial not only in patients with constipation due to lower adverse reactions (lower values of Cmax) and longer and more stable effects (higher final concentrations and more time in body). Levodopa combined with carbidopa constitutes one of the most frequent medications in the treatment of Parkinson's disease; the study was therefore extended by the same group of scientists to probe such interactions in the presence of carbidopa. The study revealed that coadministration in this case led to reduction of efficacy.		Garcia et al. 2005; Fernandez et al. 2005
		43. Munoz et al. from the Department of Cellular Biology, Physiology and Immunology, University of Barcelona, described a case of bronchial asthma due to occupational exposure from inhalation of Aescin. A 57-year-old worker employed in the pharmaceutical industry developed the disease while working with products, for example, *Plantago ovata* and Aescin (an active ingredient of this plant with anti-inflammatory and venotoxic properties). Occupational exposure was confirmed by several laboratory tests including a specific inhalation challenge with Aescin. The mechanism involved is unknown. An irritative mechanism, secondary to chronic low-level exposure, cannot be ruled out.	Aescin in *Plantago ovata*	Munoz et al. 2006

(Continued)

TABLE 3.1 (CONTINUED)
Global Picture: 59 Examples of the Adverse Effects of Single-Component and Multicomponent Herbal Drugs from 27 Countries

S. No.	Country	Toxic and Adverse Side Effects or Interactions	Toxic or Active Principle	References
23.	Spain	44. Alvarez-Perea et al. from the University of Gregor Muranon, Madrid, reported a case of urticaria due to *Aloe vera*. The affected patient (a 57-year-old woman) applied the peeled-off rind of the plant on a wound on her left eyebrow. Four hours later, she woke up with intense generalized pruritus; hive in her arms, thorax, and abdomen; erythema; and edema in her hands, feet, and eyebrows. She was treated in the emergency department with methyl-prednisolone and dex-chlorpheniramine and recovered after two hours.		Alvarez-Perea et al. 2010
24.	Sri Lanka	45. Arseculeratne et al. from the University of Colombo extensively reviewed the medicinal plants of Sri Lanka for their potential therapeutic value and safety aspects. They stated that no herb–herb or herb–drug interactions have been reported in literature for the plant *Withania somnifera*. While some toxicity has been observed in experimental animals, it was not serious, and the plant is quite safe at reasonable doses in human beings.		Arseculeratne et al. 1985
25.	Turkey	46. Scientists from Haydorposa Numune Education and Research Hospital, Istanbul, reported a case of multiorgan toxicity including renal failure due to rhabdomyolysis, acute hepatotoxicity accompanied by angioedema in a Turkish male patient after ingestion of mixed Chinese herbs. The authors suggested the involvement of hypersensitivity reactions as a possible mechanism for pathogenesis.	–	Berrin et al. 2006
26.	United Kingdom	47. Raman and Lau from King's College of London reported phytochemistry and toxic effects of *Momordica charantia* L. They stated that while a wide range of beneficial effects of unripe fruits seeds, aerial parts, and purified compounds have been reported, some of these actions are observed at very high doses. The principal toxicity of this herbal drug is on the liver and reproductive system in animals, but such adverse effects have not been reported in humans despite the widespread use of the fruit as a kitchen vegetable and for medicinal purposes.		Raman and Lau 1996

(Continued)

TABLE 3.1 (CONTINUED)
Global Picture: 59 Examples of the Adverse Effects of Single-Component and Multicomponent Herbal Drugs from 27 Countries

S. No.	Country	Toxic and Adverse Side Effects or Interactions	Toxic or Active Principle	References
		48. Kean et al. from the Department of Dermatology and Clinical Biochemistry, King's College Hospital, London, analyzed 11 Chinese herbal creams that were collected from patients who were attending general and pediatric dermatology outpatient clinics. A total of 10 patients submitted the creams for analysis (7 children and 3 adults). One adult was using two creams simultaneously, giving a total of 11 samples for analysis. Eight creams were found to contain dexamethasone at a mean concentration of 456 μg^{-1} (range: 64–1500 μg^{-1}). The analysis was done using high-resolution gas chromatography and mass spectrometry. The peaks compared well with known standard steroid peaks. The researchers warned against inadvertent use of topical steroids because they can cause severe exacerbation of *Eczema herpeticum*. These creams were applied on sensitive skin areas such as the face and flexures and in children. They recommended strict regulations to prevent illegal and inappropriate prescribing of potent steroids. Labels of such Chinese herbal preparations may not indicate the content.	Steroids	Kean et al. 1999
		49. Wilcox et al. from the Research Initiative on Traditional and Antimalarial Methods (RITAM), Oxford, designed a score to evaluate over 1200 plant species used for the treatment of malaria worldwide with the aim to prioritize plants for in-depth laboratory studies and possibly clinical studies. RITAM was founded in 1999 with the objective to review the current knowledge on the subject, select plants with promise, and avoid replication of research. It was found that plants widely used as antimalarial by traditional healers were significantly more active *in vitro* and *in vivo* against *Plasmodium* than plants that are not widely used for the treatment of malaria. The authors claimed that this approach has proved to work well. They concluded that the RITAM score (for efficacy and safety components) is useful in selection of plants for priority research, but the validity is limited by (1) a small number of clinical studies and (2) heterogeneity of patients who are included.		Wilcox et al. 2011

(*Continued*)

TABLE 3.1 (CONTINUED)
Global Picture: 59 Examples of the Adverse Effects of Single-Component and Multicomponent Herbal Drugs from 27 Countries

S. No.	Country	Toxic and Adverse Side Effects or Interactions	Toxic or Active Principle	References
26.	United Kingdom	50. A global rise in the use of traditional Chinese medicine (TCM) and *Chinese Materia Medica* (CMM) medicinal product is the cause for concern for their evidence base, safety or possible toxicity, and questionable quality. There is paucity of knowledge regarding the range plant species to produce *CMM* in Europe and even in China. Researchers from the University of Reading School of Pharmacy, Berkshire, made a first attempt to identify and compare the most important *CMM* that is used in both regions to begin the process of assessing risks to public health and possible future benefits. Data were collected from a convenient sample of TCM practitioners in the European Union (EU) and maintained in China using a paper or online survey. From a total of 747 usable questionnaires (China: 420, EU: 372), it was found that a. TCM was generally more commonly used for obstetric or gynecological and dermatological conditions in the EU compared to gastrointestinal and neurological diseases in China. b. Disorders treated by Chinese practitioners were more varied and often for more serious conditions than those that are treated by their EU counterparts. c. The range of materials used in China was wider. d. The potential for toxicity was not high in either region; though greater in China due to the use of more potent CMM, the incidence of side effects was perceived to be higher by EU practitioners. e. The main cause of concern is likely to be interaction with prescribed medication, especially in CNS and cardiovascular conditions, where drug interactions have previously been reported more frequently. This would be currently more applicable in China than in EU countries.		Williamson et al. 2013

(Continued)

TABLE 3.1 (CONTINUED)
Global Picture: 59 Examples of the Adverse Effects of Single-Component and Multicomponent Herbal Drugs from 27 Countries

S. No.	Country	Toxic and Adverse Side Effects or Interactions	Toxic or Active Principle	References
27.	United States	51. Kendler from Manhattan College, New York, reviewed the relationship of Garlic (*A. sativum*) and Onion (*Allium cepa*) to cardiovascular diseases (CVDs). He stated that (1) while many of the studies have serious methodological shortcomings, there is some evidence to suggest that use of certain formulations of garlic and/or onion is accompanied by favorable effects on risk factors (hyperlipidemia, hypertension, hyperglycemia, and possibly on platelet aggregation and blood fibrinolytic activity), and (2) the possibility of toxicity (from acute and chronic ingestion of large amounts of these plants or their extracts) is unresolved. Additional research in this area is recommended before the role of these plants in the prevention and control of CVD is understood and can be realized.		Kendler 1987
		52. Researchers from Greenwich Hospital, Peak Wellness, Greenwich, National Drug Research Institute, New York, and TSI University, Houston, Texas, investigated the effects of *Citrus aurantium* extract (CE), caffeine (CF), and St. John's Wort (JW) on body fat loss, lipid levels, and mood states in overweight healthy adults. Double-marked, randomized, placebo-controlled studies on 23 subjects with high body mass index were assigned to three groups. Group A received CE 975 mg, CF 528 mg, and JW 900 mg daily; Group B received maltodextrin placebo; and Group C received nothing and served as a control group. For six weeks, the subjects were instructed to follow a 1800 kcal/day Step 1 diet recommended by the American Heart Association and perform three days/week circuit training exercise under the supervision of an exercise physiologist. During the exercise, subjects achieved approximately 70% of age-predicted heart rate. Group A (treated) showed significant weight loss (1.4 kg) and body fat loss (mean change: 2.9%) versus group B (placebo) and group C (control). Changes in blood pressure, heart rate, electrocardiogram (ECG), cholesterol or triglyceride levels, mood states, vigor, or fatigue were not found to be statistically significant.		Colker et al. 1999

(*Continued*)

TABLE 3.1 (CONTINUED)

Global Picture: 59 Examples of the Adverse Effects of Single-Component and Multicomponent Herbal Drugs from 27 Countries

S. No.	Country	Toxic and Adverse Side Effects or Interactions	Toxic or Active Principle	References
27.	United States	It was concluded that the combination of CE, CF, and JW is safe and effective when combined with mild calorie restriction and exercise for promoting both body weight and fat loss in healthy overweight subjects.		
		53. The task of caring perioperatively for patients who use herbal medications is quite challenging. The limited evidence-based information about the safety and efficacy of such drugs, the absence of standard regulatory mechanism for herbal medicine approval and surveillance and improper patient assumptions about their safety are important medical issues. Although there has been initiation of herbal medicine into the curricula at several institutions in the United States, many practicing physicians remain unaware of the potential perioperative complications that are associated with the use of herbal medication. Ang-Lee et al. from the University of Chicago addressed these issues in an extensive review (covering database searches for published papers between 1966 and 2000). They selected studies, case reports, and reviews on pharmacology and safety of eight commonly used herbal medications for the purpose. These are		Ang-Lee et al. 2001
		a. *Allium sativum* (Garlic): It is one of the most extensively researched plant drugs with antihypertensive, hypocholesterolemic, hypolipidemic, antithrombotic, and atherosclerosis-protecting properties. Active principle is sulfur-containing Allicin and its transformation products. Commercial garlic preparations may be standardized to a fixed Allin and Allicin content: i. Garlic prevents platelet aggregation in a dose-dependent manner by inhibiting platelet-activating factors. ii. Ajoene may cause irreversible toxication.	Ajoene	

(*Continued*)

TABLE 3.1 (CONTINUED)
Global Picture: 59 Examples of the Adverse Effects of Single-Component and Multicomponent Herbal Drugs from 27 Countries

S. No.	Country	Toxic and Adverse Side Effects or Interactions	Toxic or Active Principle	References
		iii. Concurrent use with other platelet inhibitors (e.g., Prostacyclin, Forskolin, Indomethacin, and Dipyridamole) increase the risk of bleeding.		
		iv. Preoperative discontinuation is recommended for at least seven days before surgery.		
		b. *Echinacea* spp. (Purple Cornflower root): It is attributed with antibacterial, antifungal, antiviral, and immunostimulant actions.		
		i. Allergic reactions (including anaphylaxis) necessitate caution for its use in patients with asthma, atopy, and allergic rhinitis.		
		ii. Decreased effect of immunosuppressant. Patients requiring immunosuppressant therapy prior to organ transplantation should be advised to avoid taking this herbal drug.		
		iii. Hepatotoxicity.		
		iv. Certain postsurgical complications (e.g., delayed wound healing, opportunistic infections) may be observed.		
		v. Should be discontinued prior to surgery in patients with liver diseases and on immunosuppressant therapy.		
		c. *Ephedra vulgaris* (Ma Huang): This plant is used to promote weight loss, increase energy, and treat respiratory diseases (e.g., asthma and bronchitis). Commercial preparations may be standardized to a fixed ephedrine content.	Ephedrine and allied alkaloids	
		i. Tachycardia and hypertension (through direct and indirect sympathomimetic effects).		
		ii. Risk of myocardial infarction, vasospasm, and thrombotic stroke.		
		iii. Interaction with halothane anesthesia may cause intraoperative ventricular arrhythmias.		
		iv. Hemodynamic instability following long-term use through depletion of endogenous catecholamines.		

(Continued)

TABLE 3.1 (CONTINUED)

Global Picture: 59 Examples of the Adverse Effects of Single-Component and Multicomponent Herbal Drugs from 27 Countries

S. No.	Country	Toxic and Adverse Side Effects or Interactions	Toxic or Active Principle	References
27.	United States	v. Interaction with MAO-inhibitors may be life threatening. vi. Discontinuation is recommended at least 24 hours before surgery.		
		d. *Ginkgo biloba* (Duck Foot Tree, Silver Apricot): It is widely used for varied effects, for example, improving cognitive functions in Alzheimer's disease and multi-infarct dementia, age-related macular degeneration, tinnitus, vertigo, altitude sickness, peripheral vascular disease, and erectile dysfunction. i. Spontaneous bleeding through inhibition of platelet-aggregating factors. ii. Pharmacokinetic data and risk of hemorrhages warrant discontinuation of this herbal drug at least 36 hours before surgery.	a. Flavonoids b. Terpenoids	
		e. *Hypericum perfortum* (St. John's Wort): It is attributed with antidepressant properties; short-term treatment is effective in cases of mild to moderate depression, but a multicenter clinical trial showed that this herbal drug is not effective in cases of major depression. Commercial preparations are often standardized to a fixed hypericin (active principle) content of 0.3%: The mechanisms involved are inhibition of serotonin, norepinephrine, and dopamine reuptake by neurons. Inhibition of MAO is insignificant. It can significantly increase the metabolism of many concomitantly used drugs that are vital to the preoperative care of certain patients. This may cause serious drug interactions: i. Induction of Cytochrome P450 enzyme affecting cyclosporine, warfarin, steroids, protease inhibitors, and possibly benzodiazepines. ii. Other herbal drugs. iii. Decrease in serum digoxin levels. iv. Discontinuation of this herbal drug is recommended at least five days before surgery. This is more important for patients who are waiting for organ transplantation and those who require anticoagulants postoperation.	a. Hypericin b. Pseudo-hyperium c. Hyperforin	

(*Continued*)

TABLE 3.1 (CONTINUED)
Global Picture: 59 Examples of the Adverse Effects of Single-Component and Multicomponent Herbal Drugs from 27 Countries

S. No.	Country	Toxic and Adverse Side Effects or Interactions	Toxic or Active Principle	References
		f. *Panax ginseng* (Ginseng; American, Chinese, and Korean varieties). This plant is reported to possess adaptogen, antistress, and tonic properties and is used in the treatment of diabetes. Commercially available preparations may be standardized to a fixed content of Ginsenoside (active principle). i. Hyperglycemia ii. Increased risk of bleeding through irreversible inhibition of platelet aggregation and reduced prothrombin time and thromboplastin time. iii. Discontinuation of this herbal drug is recommended at least seven days before surgery.	Ginsenosides (a group of steroidal saponins)	
		g. *Piper methysticum* (Kava): Kava is attributed with anxiolytic, hypnotic, sedative, neuroleptic, and antiepileptic properties. The effects are mediated through potentiation of gamma aminobutyric acid (GABA) – inhibitory neurons. i. Kavalactones increase barbiturate-induced sleeping time in laboratory animals. Interactions have been reported with several CNS drugs; potential to enhance the effects of tranquillizers, sedatives, and anesthetics. ii. A case of coma attributed to an Alprazolam–Kava interaction has been reported. iii. Long-term intake or abuse may cause addiction, tolerance, and acute withdrawal symptoms. iv. Overdose or chronic intake may cause Kava dermatopathy with reversible scaly cutaneous eruptions. v. Pharmacokinetic data and possibly of potentiation of anesthetics suggest that patients should be advised to discontinue its use at least 24 hours before surgery.	Kavalactones	

(Continued)

TABLE 3.1 (CONTINUED)
Global Picture: 59 Examples of the Adverse Effects of Single-Component and Multicomponent Herbal Drugs from 27 Countries

S. No.	Country	Toxic and Adverse Side Effects or Interactions	Toxic or Active Principle	References
27.	United States	h. *Valeriana officinalis* (Valerian, Garden Heliotrope): Most of the herbal sleep aid drugs to treat insomnia contain Valerian. Sesquiterpenes are the primary source of its pharmacological effects. Actions are medicated through GABA neuron–transmission and receptor function. Commercial preparations may be standardized to known Valerinic acid content. i. Barbiturate-induced sleeping time is potentiated in experimental animals. ii. Interactions with CNS drugs. Potential to exaggerate the sedative effects of anesthetics. iii. A case of acute (Benzodiazepine-like) withdrawal syndrome has been reported with delirium and cardiac complications following surgery. The symptoms were attenuated by benzodiazepine administration. iv. Potential to increase anesthetic requirement in chronic users. v. Pharmacokinetic data available are inadequate to discontinue its preoperative use. Rather, caution is needed regarding its abrupt withdrawal. The researchers warned that patients may not volunteer the information regarding herbal drug intake and advised the physicians to put in extra effort to get such information to enable them to recognize the potential problems that may be serious. Preoperative withdrawal was recommended in 6 out of 8 of the drugs studied.	Sesquiterpenes	
		54. Some herbal drugs used as alternative medicines may be hepatotoxic. Many conventional drugs (e.g., Isoniazid and valproic acid) are also known to exhibit such effects. Hailer et al. from the Department of Medicine, Division of Clinical Pharmacology, University of California, San Francisco, and California Poison Control System, Schools of Pharmacy and Medicine, stressed the importance of making a diagnosis of toxic hepatitis linked to herbal drugs.	l-tetrahydro palmatine in *Shu Ling* (Chinese herbal drug)	Hailer et al. 2002

(Continued)

TABLE 3.1 (CONTINUED)
Global Picture: 59 Examples of the Adverse Effects of Single-Component and Multicomponent Herbal Drugs from 27 Countries

S. No.	Country	Toxic and Adverse Side Effects or Interactions	Toxic or Active Principle	References
		a. *Plants studied*: Ten plants: (1) *Atractylis gummifera*, (2) *Cassia angustifolia* (Senna), (3) *Chenopodium majus* (Greater Celodine), (4) *Crotolaria* spp. (Bush Tea), (5) *Ephedra* spp. (Ma Huang), (6) *Laurea tridentata* (Chappral), (7) *Senecio longifolius*, (8) *Tecurium chamaedrys* (Germander), (9) *Tussiligo farfra* (Colt's Foot), and (10) *Viscum album* (Mistletoe) b. *Formulation studies*: Five combinations: (1) Asian herbs used for eczema and psoriasis, for example, *Dictamus dosycarpus*, *Glycyrrhiza* spp., *Laphatherum* spp., *Paeonia* spp., and *Rehmania glutimosa*; (2) Jin Bu Huang containing three plants: *Corydalis* spp., *Lycopodium serratum*, and *Stephania* spp.; (3) Pennyroyal containing two plants: *Hedema pulegoides* and *Mentha pulegium*; (4) three species of *Symphytum* used as Comfrey: *S. asperum*, *S. officinale*, or *S. uplandicum*; and (5) Skullcap–Valerian combination comprising *Scutelleria laterniflora* and *Valeriana officinalis*. c. *Literature review* (1966–2002): (1) Helped in the identification of 72 cases of hepatotoxicity that are associated with 15 plant products (10 plants and 5 combinations or formulations). (2) The maximum number of cases were reported for plant 6 (16 cases), followed by plants 3 and 8 (10 cases for each plant). (3) Only 1–2 cases were reported for plants 1, 2, 5, 7, 9, and 10. (4) No case was reported for plant 4 in the United States. Epidermics have, however, been reported for bush tea in developing countries that were attributed to contamination. (5) Symptoms or lesions and the mechanism involved (wherever known) were tabulated for the plants studied. Plants 4, 7, and 9 caused veno-occlusive disease, portal hypertension, centrilobular necrosis, and hepatocellular carcinoma. Centrilobular necrosis was found to be the predominant feature of hepatic injury for plant 2.		

(Continued)

TABLE 3.1 (CONTINUED)
Global Picture: 59 Examples of the Adverse Effects of Single-Component and Multicomponent Herbal Drugs from 27 Countries

S. No.	Country	Toxic and Adverse Side Effects or Interactions	Toxic or Active Principle	References
27.	United States	Hepatocytic necrosis and choleostasis were observed with plant 6. Direct hepatotoxic action was described or suspected for five herbal products (1, 2, 4, 7, and 9). While the mechanisms involved for the remaining products (3, 5, 6, 8, and 10) are not known, the possibility of idiosyncratic and immune-mediated mechanisms was suggested for plants 3 and 8, respectively. d. *Case studies*: The authors investigated two contrasting case studies reported to the California Poison Control System to illustrate the importance and challenges in diagnosing herb-related liver injury. *Patient 1*: A previously healthy 42-year-old woman, who was taking three herbal homeopathic preparations (combination ii), developed nausea, abdominal pain, and jaundice. Acute hepatitis was diagnosed by the primary healthcare physician and confirmed on laboratory investigations. Though the cause of disease could not be identified, the alert physician suspected that herbal products might be responsible and advised the patient to stop using these products. Chemical analysis of the suspected product revealed l-tetrahydropalmatine to be the hepatotoxic ingredient that is present in *Shu-Ling*. The product lot was confiscated from the importer or seller. A public health warning was issued by the California Department of Health Sciences. The poison index database found *Jin Bu Huan* as a synonym for *Shu-Ling*. The former is banned from import to the United States. It was cleared by the U.S. custom authorities because only its Chinese name was mentioned on the shipping invoice. The other two herbal ingredients of the combination contained no hepatotoxin. The patient recovered completely after a six-week intensive hospital care.		

(Continued)

TABLE 3.1 (CONTINUED)
Global Picture: 59 Examples of the Adverse Effects of Single-Component and Multicomponent Herbal Drugs from 27 Countries

S. No.	Country	Toxic and Adverse Side Effects or Interactions	Toxic or Active Principle	References
		Patient 2: A 39-year-old female office worker began taking herbal products paradoxically for "detoxication." When she was a teenager, she suffered from hepatitis that is associated with infectious mononucleosis. She did not seek medical care but rather sought self-help from a herbal therapy book; solicited advice from a local nutrition store; took several herbal products; and was reluctant to disclose or recall the number, amount, and duration of product intake. She was found in a serious condition at home and was admitted to the hospital with symptoms of mild jaundice, sclerus, icteras, hypertension, and respiratory problems, which progressed to hepatic failure, hepatorenal syndrome, and encephalopathy rapidly. These were attributed to intake of a "liver tonic" (Plant 6 Chappral) 500 mg daily for two days before hospitalization. The patient underwent orthotopic liver transplantation. Postoperative course was complicated, but after 21 days of intensive care, she eventually recovered and was discharged from the hospital. e. *Concluding remarks*: Both cases had a specific medical condition that warranted healthcare consultation and intervention. Early diagnosis by the alert physician, patient cooperation, and systematic hospital care in the first care lead to quick recovery. Poor history, multiple product use, self-help, and reluctance to seek medical advice or treatment in the second case resulted in serious disease necessitating liver transplantation, which could have been avoided.		

(Continued)

TABLE 3.1 (CONTINUED)
Global Picture: 59 Examples of the Adverse Effects of Single-Component and Multicomponent Herbal Drugs from 27 Countries

S. No.	Country	Toxic and Adverse Side Effects or Interactions	Toxic or Active Principle	References
27.	United States	55. Setty and Sigal from Massachusetts General Hospital, Boston Johnson University Hospital and Pharmaceutical Research Institute, Bristol-Mayers Squibb, Princeton, reviewed herbal medicine commonly used in the practice of rheumatology. Investigations on the mechanism and potential of complementary and alternative health (CAH) therapies are still in their infancy, and many studies done to date are scientifically flawed. Further systematic and scientific studies on this issue are warranted to validate or refute the clinical claims that were made by CAH therapies. An understanding of the mechanisms involved should enable the physicians to advise effectively their proper and improper use, prevent adverse drug interactions, and anticipate toxicities.		Setty and Sigal 2005
		56. Newman and Cragg from the Natural Products Branch, Developmental Therapeutic Program of the National Cancer Institute, Frederick, Maryland, presented an extensive review of natural products as the source of new drugs for the last 25.5 years (January 1981 to June 2006). a. Classification of chemical entities as a source of drugs: **Category (Origin of Drug)** — **Number** i. *Biological* (B): Usually large (>45 residues) peptide or protein either isolated from an organism or cell line or produced by biotechnology means in a surrogate host — 124 ii. *Natural products* (N) — 43 iii. *Natural product derivative* (ND): Usually a semisynthetic modification — 232		Newman and Cragg 2007

(*Continued*)

TABLE 3.1 (CONTINUED)
Global Picture: 59 Examples of the Adverse Effects of Single-Component and Multicomponent Herbal Drugs from 27 Countries

S. No.	Country	Toxic and Adverse Side Effects or Interactions	Toxic or Active Principle	References
		iv. *Synthetic (S)*: Totally synthetic, often founded by random screening or modification of an existing agent 310 v. *Synthetic (S*)*: Made by total synthesis, but pharmacophore is or was from a natural product 47 vi. *Vaccine (V)* 39 vii. *Natural product mimic (S-NM)* 108 viii. *Natural product mimic (S*-NM)* 107 **Total 100** Medical indications for 60 diseases were also tabulated for these categories. Antibacterial and anticancer drugs topped the list with 109 and 100 chemical entities, respectively. b. *Comments*: From the data presented, the utility of natural products as sources of novel structures, but not necessarily the final drug entity, is still alive and well. In the area of cancer, over the time frame from around the 1940s to 2006, of the 155 small molecules, 73% are other than synthetic, with 47% actually being natural products or directly derived from there. In other areas, the influence of natural product structures is quite marked with anti-infective ranking high in drug development from natural products and their derivatives.		

(Continued)

TABLE 3.1 (CONTINUED)
Global Picture: 59 Examples of the Adverse Effects of Single-Component and Multicomponent Herbal Drugs from 27 Countries

S. No.	Country	Toxic and Adverse Side Effects or Interactions	Toxic or Active Principle	References
27.	United States	Although combinatorial chemistry techniques have succeeded as methods of optimizing structures and have, in fact, been used in optimal use of many recently approved agents, the authors were able to identify only one *de novo* combinatorial compound approved as a drug in this 25+ year time frame. They sought the reader's attention to this rapidly evolving recognition that a significant number of natural product drugs or leads are actually produced by microbes and/or microbial interaction with the "host from where it was isolated."		
		57. Booth et al. from West Lake Partners, Chester, studied a natural herbal drug supplement (NT), containing Rhubarb, Astralagus, Red Sage, Ginger, and Turmeric, for reproductive toxicity when given in combination with Gallic acid (GA). The study was carried out in 50 pregnant rats (NT 80%, GA 20% mixture, 21.6, 215, 430, and 860 mg kg^{-1} per day by daily gavage from days 6–20 of gestation). All 50 rats had one or more live fetuses and survived till they were sacrificed; the combination elicited no adverse effects on food intake, corpora lutea, implantations, number and size of fetuses up to 430 mg kg^{-1} dose level. This herbal dietary supplement, taken for the treatment of obesity, appears to be quite safe up to this dose level. A higher dose (860 mg kg^{-1}) exhibited some reprotoxicity.		Booth et al. 2010

(*Continued*)

TABLE 3.1 (CONTINUED)
Global Picture: 59 Examples of the Adverse Effects of Single-Component and Multicomponent Herbal Drugs from 27 Countries

S. No.	Country	Toxic and Adverse Side Effects or Interactions	Toxic or Active Principle	References
		58. *Glycyrhhiza* (Licorice) extract has always been recognized as a sweetener and thirst quencher. It is also used widely for medicinal purposes. Its value is overrated by many who consume significant amounts and are prone to its complications. Glycyrrhetic acid, the active metabolite, inhibits 11-β-hydroxysteroid dehydrogenase enzyme type 2 with a resultant cortisol-induced mineralocorticoid effect and aldosterone-like action. This is the basis for understanding its health benefits and a wide spectrum of adverse effects. Researchers from Mercy Hospital and Medical Center, Chicago, presented a comprehensive review on this herb. It highlights the importance of investigating worldwide usage of herbal remedies on cultural and habitual bases rather than reliable scientific evidence. Licorice is a FDA-approved food supplement that is used in many products without precise regulations to prevent toxicity. The authors stressed the need to send a warning message to prevent its abuse. Increased public awareness is warranted through media (TV commercials, newspapers, magazines, Internet sites) and product labels regarding upper limit of ingestion and health hazards that are associated with excess intake. This will enable the physicians to advise and help their patients.	Glycyrhhetic acid	Omar et al. 2012

(*Continued*)

TABLE 3.1 (CONTINUED)
Global Picture: 59 Examples of the Adverse Effects of Single-Component and Multicomponent Herbal Drugs from 27 Countries

S. No.	Country	Toxic and Adverse Side Effects or Interactions	Toxic or Active Principle	References
27.	United States	59. Due to the lack of objective tests to diagnose drug-induced liver injury, casualty assessment on such issues is a matter of debate. Hayashi et al. from the Division of Gastroenterology and Hepatology of University of California described five categories for Drug-Induced Liver Injury Network (DILIN) representing levels of likelihood. Adjudication is based on retrospective assessment of gathered data that includes prospective follow-up information. One hundred randomly selected DILIN cases were reassessed using the same processes for initial assessment but by three different reviewers in 92% of cases. The results are as follows: a. The median time between assessments was 938 days (range: 140–2352). b. About 31% cases involved more than one agent. c. The weighted kappa statistics for overall case and individual agent category agreement were 0.60 (0.50–0.71) and 0.60 (0.52–0.68), respectively. d. Overall case adjudications of each other: 93% within one category, 5% differed by two categories, and 2% differed by three categories. e. About 14% crossed the 50% threshold of likelihood due to competing diagnoses or atypical timing between drug exposure and injury. It was concluded that the DILIN expert opinion causality assessment method has moderate interobserver reliability but very good agreement within one category. A small, but important, proportion of cases could not be reliably diagnosed as ≥50% likely.	–	Hayashi et al. 2014

REFERENCES

Abraham, K., Wohrlin, F., Lindtner, O., Heinmeyer, G., and Lampen, A. 2010. Toxicology and risk assessment of coumarin: Focus on human data. *Mol Nutr Food Res* 54: 228–238.

Adedapo, A.A., Omoloya, O.A., and Ohore, O.G. 2007. Studies on the toxicity of an aqueous extract of the leaves of *Abrus precatorius* in rats. *Onderstepoort J Vet Res* 74: 31–36.

Alvarez-Perea, A., Garcia, A.P., Hernandez, A.L.R., De Barrio, M., and Baeza, M.L. 2010. Urticaria due to aloe vera: A new sensitizer? *Ann Allergy Asthma Immunol* 105: 404–405.

Ang-Lee, M.K., Moss, J., and Yuan, C.S. 2001. Herbal medicines and perioperative care. *JAMA* 286: 208–216.

Ansah, C. and Mensah, K.B. 2013. A review of the anticancer potential of the antimalarial herbal *Cryptolepis sanguinolenta* and its major alkaloid cryptolepine. *Ghana Med J* 47: 137–147.

Arseculeratne, S.N., Gunatilaka, A.A., and Panshokke, R.G. 1985. Studies on medicinal plants of Sri Lanka Part 14. Toxicity of some traditional medicinal herbs. *J Ethnopharmacol* 13: 323–335.

Ashafa, A.O.T., Orekoya, L.O., and Yakubu, M.F. 2012. Toxicity profile of ethanolic extract of *Azadirachta indica* stem bark in Wistar rats. *Asian Pac J Trop Med* 2: 811–817.

Berrin, Y., Ali, O., Umut, S., Meltem, E., Murat, B., and Barut, Y. 2006. Multi-organ toxicity following ingestion of mixed herbal preparations: An unusual but dangerous adverse effect of phytotherapy. *Eur J Intern Med* 17: 130–132.

Bnouham, M., Merhfourl, F.Z., Mostafa, E., Legssyer, A., Hassane, M., Lamnaouer, D., and Abderrahim, Z. 2006. Toxic effects of some medicinal plants used in Moroccan traditional medicine. *Moroccan J Biol* 2–3: 21–30.

Boeke, S.J., Boersma, M.G., Alink, G.M., Van Loon, J.J.A., Van Huis, A., Dicke, M., and Rietiene, I.M.C.M. 2004. Safety evaluation of neem (*Azadirachta indica*)-derived pesticides. *J Ethnopharmacol* 44: 25–41.

Booth, A., Amen, R.J., Scott, M., and Greenway, F.L. 2010. Oral dose-ranging developmental toxicity on an herbal supplement (NT) and gallic acid in rats. *Adv Ther* 27: 250–255.

Bugyei, K.A., Boye, G.L., and Addy, M.E. 2010. Clinical efficacy of a tea-bag formulation of *Cryptolepis sanguinolenta* root in the treatment of acute uncomplicated falciparum malaria. *Ghana Med J* 44: 3–9.

But, P.P.M. 1993. Need for correct identification of herbs in herbal poisoning. *Lancet* 340: 637.

Chan, T.Y., Tam, H.P., Lai, C.K., and Chan, A.Y.W. 2005. A multidisciplinary approach to the toxicologic problems associated with the use of herbal medicines. *Ther Drug Monit* 27: 53–57.

Chen, S. and Viera, A. 2010. A meta-analysis of medicinal plants to assess the evidence for toxicity. *Interdisc Toxicol* 3: 82–85.

China Pharmacopoeia Committee. 2010. *Pharmacopoeia of the People's Republic of China.* 36: 177–220.

Chinese Pharmacopoeia. 2010. *Pharmacopoeia of the People's Republic of China* (English edition), 9th edition, Volumes I to III.

Colker, C.M., Kaiman, D.S., Torina, G.C., and Perlis, T. 1999. Effects of *Citrus aurantium* extract, caffeine, and St. John's wort on body fat loss, lipid levels, and mood states in overweight healthy adults. *Curr Ther Res* 60: 145–153.

Coopman, V., De Leeuw, M., Cordonnier, J., and Jacobs, W. 2009. Suicidal death after injection of a castor bean extract (*Ricinus communis* L.). *Forensic Sci Int* 189: e13–e20.

Cravotto, G., Boffa, L., Genzine, L., and Garella, D. 2010. Phyto-therapeutics: An evaluation of potential 1000 plants. *J Clin Pharm Ther* 35: 11–48.

Creemers, L., Driessche, M.V., Moens, M., Olmen, A.V., Verschaeren, J., Syen, M.T., Debmet, K., and Moerman, J. 2007. Safety of alternative medicines reconsidered: Lead-induced anemia caused by an Indian Ayurvedic formulation. *Arch Clinica Biologica* 63: 548–551.

Degraeve, N. 1978. Genetic and related effects of *Vinca rosea* alkaloids. *Mutat Res.* 55: 31–42.

Ekor, M. 2013. The growing use of herbal medicines: Issues relating to adverse reactions and challenges in monitoring safety. *Front Pharmacol* 4: 1–10.

Faghihi, G. and Radan, M. 2011. Side effects of herbal drugs in dermatology. *J Cosmetics, Dermatol Sci App* 1: 1–3.

Fernandez, N., Carriedo, D., Sierra, M., Diez, M.J., Sahagun, A., Calle, A., Gonzalez, A., and Garcia, J.J. 2005. Hydrosoluble fiber (*Plantago ovata* husk) and levodopa II: Experimental study of the pharmacokinetic interaction in the presence of Carbidopa. *Eur Neuropsychopharmacol* 15: 505–509.

Garcia, J.J., Fernandez, N., Carriedo, D., Diez, M.J., Sahagun, A., Gonzalez, A., Calle, A., and Sierra, M. 2005. Hydrosoluble fiber (*Plantago ovata* husk) and levodopa I: Experimental study of the pharmacokinetic interaction. *Eur Neuropsychopharmacol* 15: 497–503.

Grieco, A., Miele, L., Pompili, M., Vecchio, F.M., Grottagliando, I., and Garburrini, G. 2009. Acute hepatitis caused by a lipid-lowering product: When alternative medicine is no alternative at all. *J Hepatol* 50: 1273–1277.

Haden, M., Marshall, D.A., and Murphy, B. 2011. Toxic levels of glycosides in herbal medication: A potential cause of hyperkalemia. *Scott Med J* 56: 236.

Hailer, C.A., Dyer, J.E., Ko, R.J., and Olson, K.R. 2002. Making a diagnosis of herbal-related toxic hepatitis. *West J Med* 176: 39–44.

Hayashi, P.H., Barnhart, H.X., Fontana, R.J., Chalasani, N., Davern, T.J., Talwalkar, J.A., Reddy, K.R., Stolz, A.A., Hoofnagle, J.H., and Rockey, D.C. 2014. Reliability of causality assessment for drug, herbal and dietary supplement hepatotoxicity in the Drug-Induced Liver Injury Network (DILIN). *Liver Int* 24.

Homma, M., Oka, K., Nitisuma, J., and Itoh, H. 1993. Pharmacokinetic evaluation of traditional Chinese herbal remedies. *Lancet* 341: 1595.

Hussin, A.H. 2001. Adverse effects of herbs and drug-herbal interactions. *Malaysian J Pharmacy* 1 (2): 39–44.

Kean, F.M., Munn, S.E., Du Vivier, A.W.P., Taylor, N.F., and Higgins, F.M. 1999. Analysis of Chinese herbal creams prescribed for dermatological conditions. *BMJ* 318: 563–564.

Kendler, B.S. 1987. Garlic (*Allium sativum*) and onion (*Allium cepa*): A review of their relationship to cardiovascular disease. *Prev Med* 16: 670–685.

Khan, M.A. and Islam, M.T. 2012. Analgesic and cytotoxic activity of *Acorus calamus L., Kigella pinnata L., Mangifera indica L.,* and *Tabernaemontana divericata. J Pharm Bioallied Sci* 4: 149–154.

Kramar, R. and Kaiser, E. 1968. Effect of imperatorin, a toxic principle from Ammi majus on energy-transfer in mitochondria. *Toxicon* 6: 145–147.

Lai, R.S., Chiang, A.A., Wu, M.T., Wang, J.S., Lai, N.S., Lu, J.Y., Ger, L.P., and Roggli, V. 1996. Outbreak of bronchiolitis obliterans associated with consumption of Sauropus androgynus in Taiwan. *Lancet* 348: 83–85.

Landau, C.A. and Sanchez, M.S. 2010. In vitro screening of plant extracts: Neurotoxic effects of the "Sirrae de Cordoba," Argentina plant extracts. *Mol Med Chem* 20: 48–52.

Li, L., Dou, L.X., Neilson, J.P., Leung, P.C., and Wang, C.C. 2012. Adverse outcomes of Chinese medicines used for threatened miscarriage: A systematic review and meta-analysis. *Hum Reprod Update* 18: 504–524.

Liu, Q., Zhou, L., Liu, L., Zhu, S., Sunnassee, A., Liang, M., Zhou, L., and Liu, Y. 2011. Seven cases of fatal aconite poisoning: Forensic experience in China. *For Sci Int* 212: e5–e9.

Luyckx, V.A. 2012. Nephrotoxicity of alternative medicine practice. *Adv Chronic Kidney D* 19: 129–141.

Melchardt, T., Magnes, T., Weiss, L., Grundbichler, M., Strasser, M., Hufnagl, C., Moik, M., Greil, R., and Egle, A. 2014. Liver toxicity during temozolomide chemotherapy caused by Chinese herbs. *BMC Complement Altern Med* 14: 115.

Misaka, S., Yatabe, J., Müller, F., Takano, K., Kawabe, K., Glaeser, H., Yatabe, M.S., Onoue, S., Werba, J.P., Watanabe, H., Yamada, S., Fromm, M.F., and Kimura, J. 2014. Green tea ingestion greatly reduces plasma concentrations of Nadolol in healthy subjects. *Clin Pharmacol Ther* 95: 432–438.

Mostera-Kara, N., Panwels, A., Pines, E., Biour, M., and Levy, V.G. 1992. Fatal hepatitis after herbal tea. *Lancet* 340: 647.

Munoz, X., Culebras, M., Cruz, M.J., and Morell, F. 2006. Occupational asthma related to aescin inhalation. *Ann Allergy Asthma Immunol* 96: 494–496.

Neergheen-Bhujan, V.S. 2013. Understanding the toxicological challenges associated with the use of herbal medicinal products in developing countries. *BioMed Res Int* 2013: 1–9.

Newman, D.J. and Cragg, G.M. 2007. Natural products as source of drugs over the last 25 years. *J Nat Prod* 70: 461–477.

Oluyemi, K.A., Okwuonu, U.C., Baxter, D.G., and Oyesola, T.O. 2007. Toxic effects of methanolic extract of *Aspilia africana* leaf on the estrus cycle and uterine tissues of Wistar rats. *Int J Morphol* 25(3): 609–614.

Omar, H.R., Komarova, I., El-Ghonemi, M., Fathy, A., Rashad, R., Abdelmalak, H.D., Yerramadha, M.R., Ali, Y., Helal, E., and Camporesi, E.M. 2012. Licorice abuse: Time to send a warning message. *Ther Adv Endocrin Metab* 3: 125–138.

Padiyara, R.S. and Khan, S.F. 2006. A review of commonly used herbal medicines. Illinois Council of Health-System Pharmacists ICHP Keeposted 9: 22–36.

Pahwa, R. and Chatterjee, V.C. 1990. The toxicity of yellow oleander (*Thevetia neriifolia* Juss) seed kernals in rats. *Vet Hum Toxicol* 32: 561–564.

Petronilho, S., Maraschin, M., Coimbra, M.A., and Rocha, S.M. 2012. In vitro and in vivo studies on natural products. A challenge for their valuation. *Ind Crop Prod* 40: 1–12.

Raman, A. and Lau, C. 1996. Anti-diabetic properties and phytochemistry of *Momordica charantia* L. *Phytomedicine* 2: 349–362.

Rates, S.M.K. 2001. Review: Plants as source of drugs. *Toxicon* 39: 603–613.

Renter, H.D. 1995. *Allium sativum* and *Allium ursinum* Part 2. Pharmacology and medical application. *Phytomedicine* 2: 73–91.

Rong, X., Peng, G., Suzuki, T., Yang, Q., Yamahara, J., and Li, Y. 2009. A 35-day gavage safety assessment of ginger in rats. *Regul Toxicol Pharmacol* 54: 118–123.

Rose, K.D., Croissant, P.D., Parliament, C.F., and Levin, M.B. 1990. Spontaneous spinal epidural hematoma with associated platelet dysfunction from excessive garlic ingestion: A case report. *Neurosurgery* 26: 880–882.

Russo, S., Mastropasqua, M., Mosetti, M., Persegani, C., and Paggi, A. 2000. Low doses of liquorice can induce hypertension encephalopathy. *Am J Nephrol* 20: 145–148.

Saidi, H. and Mofidi, M. 2009. Case report: Toxic effect of *Xanthium strumarrium* herbal medicinal preparation. *EXCLI J* 8: 115–117.

Schoepfer, A.M., Engel, A., Fattinger, K., Marbet, U.A., Criblez, D., Reichen, J., Zimmermann, A., and Oneta, C.M. 2007. Herbal does not mean innocuous: Ten cases of severe hepatotoxicity associated with dietary supplements from Herbalife products. *J Hepatol* 47: 521–526.

Setty, A.R. and Sigal, L.H. 2005. Herbal medicines commonly used in the practice of rheumatology. Mechanism of action, efficacy and side effects. *Semin Arthritis Rheum* 34: 773–784.

Sheng, Y., Ma, Y., Deng, Z., Wang, Z., and Ji, L. 2014. Cytokines as potential biomarkers of liver toxicity induced by *Dioscorea bulbifera* L. *Biosci Trends* 8: 32–37.

Strippoli, S., Lorusso, V., Albano, A., and Guida, M. 2013. Herbal-drug interaction induced rhabdomyolysis in a liposarcoma patient receiving trabectedin. *BMC Complement Altern Med* 30; 13: 199.

Talasaz, A.H., Abbasi, M.R., Abkhiz, S., and Dasti-Khavidakis. 2010. *Tribulus terrestris*-induced severe nepthrotoxicity in young healthy male. *Nephrol Dial Transplant* 25: 3792–3793.

Theoharides, T.C. 1997. Sudden death of a healthy college student related to ephedrine toxicity from a ma huang-containing drink. *J Clin Psychopharmacol* 17: 437–439.

Uddin, S.J., Grice, I.D., and Tirolongo, E. 2011. Cytotoxic effects of Bangladeshi medicinal plant extracts. *Evid Based Complementary and Altern Med* 2011: 1–7.

Vermaak, I., Viljoen, A., and Lindström, S.W. 2013. Hyperspectral imaging in the quality control of herbal medicines—The case of neurotoxic Japanese star anise. *J Pharm Biomed Anal* 75: 207–13.

Wang, C.C., Lu, L., Tang, L.Y., and Leung, P.C. 2012. Safety evaluation of commonly used Chinese herbal medicines during pregnancy in mice. *Human Reprod* 27: 2448–2456.

Wilcox, M., Benoit-Vical, F., Fowler, D., Boirdys, G., Burford, G., Giani, S., Gruzios, R., Houghton, P., Randrianarivelojosia, M., and Rasonaivo, P. 2011. Do ethanobotanical and laboratory data predict clinical safety and efficacy of antimalarial plants? *Malaria J 10* (Suppl.1): 57.

Wilkinson, J.M. 2000. Effect of ginger tea on the fetal development of Sprague-Dawley rats. *Reprod Toxicol* 14: 1507–1512.

Williamson, E.M., Lorenc, A., Booker, A., and Robinson, N. 2013. The rise of traditional Chinese medicine and its materia medica: A comparison of the frequency and safety of materials and species used in Europe and China. *J Ethnopharmacol* 149: 453–462.

4 The Indian Scene

4.1 INTRODUCTION

The idea that all herbal drugs are safe is a myth. There are many poisonous plants used in the traditional system of medicine that have been listed as poisonous in ancient texts and in Schedule E1 of the Drugs and Cosmetics Act (1940) (Drugs and Cosmetics Act 1940 and Rules 1945; Katiyar 2012). Twenty-five such plants are listed in Table 4.1.

Recently, Chandra Shekhar et al. (2012) reviewed the useful and harmful effects of plant toxins. They stated that a plant grows in a competitive environment. It is continuously threatened by other plants and animals. In order to survive, it develops toxic chemicals during the evolutionary process. Plant toxins (phenolics, terpenes, steroids, cytogenetic compounds, alkaloids, etc.) act on all parts of the human body with some common symptoms such as nausea and vomiting. The authors tabulated 107 plant toxins with both useful and harmful effects on the body. The intensity of such effects depends on the form of exposure and the concentration.

The discovery of reserpine from the Ayurvedic plant *Rauwolfia serpentina* (Sarapgandha) in 1952 was lauded as a wonder drug for use in hypertension and psychosis. Within the next 10 years, its side effects (depression and suicidal tendencies) came to light, and it was replaced by safer and equally therapeutically effective phenothiazines. Until that time, it was firmly believed that herbal drugs were safer versus synthetic drugs. Later, a number of plants used as herbal teas in the United Kingdom, when clinically evaluated, were found to have serious enough side effects demanding their withdrawal from the market (Dandiya 1999). George (2011) presented a critical overview of concerns regarding the safety and toxicity of medicinal plants, considered safe over many decades, and remarked that evidence suggests otherwise. He stated the importance of a clear understanding of the pros and cons for the use of Aconite, Alfa Alfa, Comfrey, Ephedra, *Ginkgo* sp., Ginseng, Isabgol, Licorice, Senna, *Silybum marianum*, and St. John's Wort.

The safety requirements for herbal formulations were discussed at a meeting organized by the Indian Society for Toxicology and Hamdard University, New Delhi (Symposium on Strategies for Safety Study Requirements for Herbal Formulations 2010). The scientific program was well timed and covered most relevant topics for utilizing the potential value of herbal products and pharmacovigilance for risk minimization and mitigation. Some highlights, relevant to this communication, are listed as follows:

a. The development of quality assurance of a medicinal plant database at the Indian Institute of Toxicology Research (IITR), Lucknow. It contains more than 33,000 entries for easy retrieval of information.

TABLE 4.1
Poisonous Plants Mentioned in Ancient Texts of Indian Systems of Medicine

Plants	Ayurveda (A), Siddha (S), and Unani Medicine (U)
Abrus precatorius	A, S
Aconitum chasmanthum	A, U
Aconitum feron	S
Adanther apavonina	S
Aloe barbadensis	S
Baliospermum montanum	A
Calotropis gigantea	A
Cannabis sativa	A, S, U
Citrullus colocynthes	S
Conium maculatum	U
Croton tiglium	
Cucumis trigonu	S
Datura metal	A, S, U
Enoecaria agallocha	S
Euphorbia grandifolia	S
Euphorbia spp.	A, S
Glorisa superba	S
Hyocymusiniber	A, U
Lawsonia inermis	S
Nerium indicum	A, S, U
Nicotiana tobacum	S
Papaver somniferum	A, S, U
Plumeria acuminata	S
Randia dumetorum	S
Strychnos nux-vomica	A, S, U

b. The establishment of a toxicity database at the Council of Scientific and Industrial Research (CSIR) Unit for Research and Development of Information Products at Pune.
c. Samples of 600 Indian medicinal plants, collected from the wild as well as various medicinal plant gardens in India by the CSIR, tested at the IITR, were found to be free from Pb, Hg, and As contaminants contrary to the claims that were made by Saper et al. (2004).
d. A study by Kakkad (2011) on more than 1800 medicinal plant samples, collected from different ecological zones in India, revealed that only 2.4% of the samples contained more than permissible levels of Pb, making 97.6% samples to be of acceptable quality for the presence of this neurotoxicant. They recommended the following plants for nontarget toxicity testing: Comfrey, Pennyroyal, Ginkgo, Ginseng, and Kava. The authors described the side effects of *Ginkgo biloba* (bleeding), St. John's Wort (gastrointestinal

disturbances, allergic reactions, fatigue, dizziness, confusion, dry mouth, and photosensitivity), Ephedra (hypertension, insomnia, arrhythmia, nervousness, tremor, headache, cerebrovascular event, myocardial infarction [MI], and renal stones), and Kava (sedation, oral and lingual dyskinesia, and exacerbation of Parkinson's disease risk). Interactions of *G. biloba* (with aspirin, warfarin, ticlopidine, dipyridimole), St. John's Wort (with antidepressants), Ephedra (with caffeine, decongestants, and stimulants), Ginseng (with warfarin), and Kava (with sedatives, sleeping pills, antipsychotics, and alcohol) were also discussed.

e. Agrawal (2010) from the Delhi Institute of Pharmaceutical Sciences and Research (DIPSAR), New Delhi, tabulated herbal drug interactions and food–drug interactions. Some examples are as follows: Hawthorn (with digitalis), Ginseng (with blood-thinning agents: garlic, ginger, and caffeine), Garlic (with anticoagulants and antidiabetic drugs), Goldenseal (with beta blockers), Feverfew (with Imitrex, an antimigraine drug), Guarana (with caffeine), Kava (with central nervous system [CNS] drugs: alcohol, barbiturates, antidepressants, and caffeine), St. John's Wort (with monoamine oxidase inhibitors), Isabgol (with antibiotics), grape juice (may increase toxicity of common drugs and enhance their action through CYP3A4 inhibition), and so on. (http://www.isrpt.co.in/Downloads/presentations_2010/day1pre14conf2010.ppt).

4.2 SINGLE PLANTS

Information on 75 commonly used medicinal or cosmetic plants is presented in the following, with scientific, local or English names; parts used, active or toxic principles (if known); principal pharmacological properties; therapeutic uses; available experimental and clinical studies; and references.

1. *Abrus precatorius* L. (Ratti, Gunja, Indian Licorice). Leaves, roots, and seeds are used for medicinal or therapeutic effects. The principal pharmacological effects are antibacterial, antifertility, and insecticidal. Other properties include diuretic, emetic, use in alopecia, and dryness of mouth (Khanna et al. 1968; Vohora 1979; Gupta and Tandon 2004). A steroidal fraction from oil is the active principle. Seeds are poisonous. Experimental studies have revealed diarrhea (within 5 hours) and death (later) following ingestion of whole, broken, and pulp of ripe seeds in (8–12 week old) white leghorn chickens (David and Peter 1970). Other toxic effects (hemagglutination, CNS depression, convulsions, respiratory and cardiac problems, enzyme inhibition, etc.) have also been reported in experimental animals (rodents, guinea pigs, rabbits, monkeys) and farm animals (cattle, sheep, horses) (Sathe et al. 1970; Bhatia 1974; Sumathi and Pattabiraman 1976; Prabhu et al. 1984; Mishra et al. 1992, 1996). LD_{50} for the aqueous extract of unprocessed and processed seeds ranged between 0.5 and 2.0 mg kg^{-1}. The values increase in processed material (with milk and *Kanji-shodit* seeds) (Singh Gautam et al. 1999). The plant has been listed as a poisonous plant

in ancient texts and Schedule E of the Drugs and Cosmetics Act of 1940. Seeds have also been reported to cause human poisoning following chewing and swallowing. Powdered seeds, when inserted into the uterus, cause abortion (Saini 2004). The toxic constituents of *A. precatorius* were reviewed by Vishwanathan and Joshi (1983). Phytotoxins and lactins in seeds have been shown to be responsible for some toxic effects (Vishwanathan and Joshi 1983; Gupta and Tandon 2004).

Please also see examples 10 and 37 under Brazil (Rates 2001) and Nigeria (Adedapo et al. 2007), respectively, in Table 3.1 in Chapter 3.

2. *Acorus calamus* L. (Bach, Ugragandha) is attributed with antipsychotic, tranquilizing, antidiarrheal, and antitubercular properties. The active principles are α- and β-Asarone. Though asarone possesses all pharmacological properties, akin to reserpine and chlorpromazine, it may cause vomiting in mammals in twice the doses that are needed for therapeutic efficacy. This side effect has prevented its use in modern medicine (Vohora 1979; Dandiya 1999; Gupta and Tandon 2004).
3. *Albizzia lebbeck* Benth (Shirish). Bark, seeds, and leaves are used for medicinal effects. These include analgesic, antiallergic, antibacterial, anticancer, antifungal, antihistaminic, anti-inflammatory, antiprotozoal, antitubercular, blood-purifying, and bronchodilatory properties. Therapeutic utility has been claimed and reported in asthma, coughs, chronic respiratory problems, and chronic skin diseases including eczema and leprosy (Hameed and Vohora 2001; Mishra 2004; WHO 2011). While experimental and clinical studies have revealed no serious toxicity in normal doses, its safety in pregnant women is not proven. However, it is safe for the infant if the nursing mother is taking this medicine. It has mild spermicidal activity, and so its use must be avoided in oligospermic persons (WHO 2011).
4. *Allium cepa* L. (Piyaz, Onion). Bulbs, leaves, and juice are used as food; eaten raw or cooked with vegetables or in soups and sauces. Onion juice is considered a panacea and attributed with varied medicinal properties and therapeutic indications: alterative, anthelmintic, antimicrobial, aphrodisiac, antispasmodic, febrifuge, stimulant, and tonic properties with utility in respiratory ailments (asthma, bronchitis, coughs, and colds), cardiac diseases (atherosclerosis, lowers blood lipids, raises high-density lipoprotein [HDL] cholesterol), infections, hay fever, and other fevers. Reports of an inhibitory effect in stomach cancer are also available. Certain precautions have been advocated: (1) It should be taken with meals, preferably raw; fried onions are difficult to digest. (2) Excessive use may promote gas and aggravate heartburn. (3) Some reports suggest that abundant use may reduce red blood cell count and hemoglobin levels (Bakhru 2001).
5. *Allium sativum* L. (Lahsun, Garlic). Bulbs are used as food in curries, ketchup, pickles, and so on. These are attributed with many medicinal properties and uses in varied diseases. These include antiarthritic, anticancer, antidiabetic, anthelmintic, anti-inflammatory, antimicrobial (antibacterial, antifungal, antiviral), antioxidant, antispasmodic, decongestant, expectorant, hypolipidemic, and immunostimulant effects with utility in diabetes,

cardiac, and respiratory ailments and for treating earaches (Vohora 1979; Hameed and Vohora 2001; Mishra 2004; Singh et al. 2009; WHO 2011). Shah and Vohora (1990) carried out experimental studies to show the enhancement of antiarthritic effects of garlic oil by boron. Lau et al. (1990) reported its preventive action against cancer. Essential oil and Allicin are considered the active principles for its varied medicinal properties.

Parmar (2005) reported an interaction of garlic with warfarin; the effect of the latter may be enhanced resulting in bleeding. Excessive use of raw garlic should be avoided. It may lead to (1) foul breath and odor from the skin, (2) a burning sensation during urination, and (3) heartburn. Such use should be avoided in pregnancy and lactation. Local application of oil is traditionally considered safe, but application on inflammatory parts may cause slight irritation and a burning sensation. While instillation of *Lahsun* oil gives relief in cases of earache, some precautionary measures have been advised: (1) The oil should not be too hot or too cold, (2) instillation should be avoided if there is discharge from the ears, and (3) it should be stopped if no relief is observed within three days of use. Medical advice should be sought in such cases (WHO 2011).

Please also see examples 17 and 53.1 under Germany (Renter 1995) and the United States (Ang-Lee et al. 2001) in Table 3.1 in Chapter 3.

6. *Aloe vera* L. (Ghritkumari, Gheekanwar, Aloe). Juice from the leaves are attributed with varied medicinal or cosmetic properties. Its principal use (as a single drug or in formulations) is in skin care or cosmetics due to its emollient, anti-inflammatory, healing (wounds, burns, radiation injury), and soothing actions. Internal administration is also prevalent through oral use. Its other claimed and reported pharmacological properties include blood-purifying, antioxidative, immunomodulatory, and laxative effects with utility in constipation, gynecological disorders, ischemic heart disease, thyroid dysfunction, sciatica, and so on (Hameed and Vohora 2001; Mishra 2004).

   Paradoxically, this drug with its principal protective use for skin care and cosmetic purpose has also been reported to be a skin sensitizer causing urticaria and other dermatological problems (Alvarez-Perea et al. 2010; Faghihi and Radan 2011).

   Please also see examples 28 under Iran (Faghihi and Radan 2011) and 44 under Spain (Alvarez-Perea et al. 2010) in Table 3.1 in Chapter 3.

7. *Amaranthus viridis* L. (Chaulai). Young roots pounded with water and sugar candy are administered to pregnant women to induce abortion (Bakhru 2001).

8. *Ammi majus* L. (Babchi, Atrilal, Bishop's weed). Its principal use is for the treatment of leucoderma or vitiligo. The drug is administered both by topical application and oral route with concurrent exposure of the affected part(s) to ultraviolet light. Extensive chemical, experimental, and clinical studies are available. The action is through furocoumarins (psoralens) in fruits, which affect tyrosinase activity that is involved in melanin pigment formation pathway. The active principle furocoumarins (psoralens) are Ammoidin, Majuddin, Iso-pimpilin, Maramerimin, Maramesminin,

Ajmajine, and Iso- and Allo-inperotonin (Chopra et al. 1980; Zafarullah and Vohora 1980; Hameed and Vohora 2001).

Furocoumarins in fruits have also been reported to cause adverse effects, for example, photosensitization, dermatitis, and minor systemic effects (nausea and giddiness) in livestock (sheep, cattle) and human subjects. Its main toxic principles are Imperotonin and Xanthotoxin (Witzel et al. 1978; Kavli and Volden 1984; Ossenkoppele et al. 1991; ICMR 2008).

9. *Amorphophallus compestris* Roxb. (Surana) is attributed with antiallergic, appetizing, carminative, and digestive properties. The plant is also used for the treatment of piles (Chopra et al. 1980; WHO 2011). Raw corn may cause itching. This can be relieved by oral intake of traditional antidotes, for example, lemon and tamarind. Some precautions and safety measures have been advised. These include (1) intake of plenty of water, cooked green vegetables, fruits, and milk, (2) avoiding a constipatory diet, hot, spicy, and fat-rich foods and straining during defecation, and (3) this drug should not be given to pregnant women and nursing mothers (WHO 2011).
10. *Anethum sowa* Roxb. ex Flem. (Shankhpuspa). Oil from the seeds is used for a variety purposes. Main pharmacological and therapeutic properties are antimicrobial (antibacterial, antifungal), antispasmodic with utility in gastrointestinal problems (antiflatulent, carminative, digestive). Other properties or indications include anthelmintic and emmenagouge effects with beneficial action against nematode infestation and cases of painful menstruation (Chopra et al. 1980; WHO 2011). Essential oil contains carvone, dihydrocarvone, limonene, dillapiol, and dillapionol. These constituents have been reported to show antifungal effects. Other pharmacological actions reported in experimental animals includes hypotensive and hypoglycemic effects. Dillapiol causes abortion and is toxic to liver and kidneys (Gupta et al. 2006a). A WHO monograph (2011) remarked that while the plant causes no serious toxic or adverse effects, certain precautions should be taken: (1) The use should be discontinued if menstrual flow increases, and symptoms of excessive thirst, burning, and giddiness are observed; (2) cold water bath and dry massage help in reducing symptoms; and (3) heavy food and stress should be avoided.
11. *Apium graveolens* L. Synonyms: *A. leptophyllatum* Pers. (Ajmod, Bekh-e-Karas). The seeds and roots are attributed with alterative, analgesic, anthelmintic, antiarthritic, anti-inflammatory, antiseptic, antispasmolytic, diuretic, and nervine sedative properties (Chopra et al. 1980; WHO 2011). While no adverse effects have been reported, some precautions and safety measures should be taken: (1) Diuretic action is elicited by irritating the renal tissue. Therefore, it should be used with caution in patients with renal disorders. (2) Arthritic persons should avoid excessive use of the affected joints. If the pain does not subside within two to three days of drug use or the joints develop effusions of fluid, medical advice should be taken. Such patients should avoid sour and cold food. Concurrent fomentation or massage with sesame oil is helpful in reducing pain. (3) The drug should not be given to pregnant women (WHO 2011).

12. *Argemone mexicana* L. (Bharbhar). Oil expressed from the seeds, if used for cooking food, causes vomiting, diarrhea, and intense body pain in humans and animals (Saini 2004).
13. *Aristolchia indica* L. (Kirmar). A decoction of the roots is used to expel roundworms. The leaves are highly purgative and cause intense diarrhea (Saini 2004).
14. *Azadirachta indica* A. Juss. Synonyms: *Melia azadirachta* L. (Neem, Margosa). Leaves, tender stems, bark, and young fruits are used for diverse medicinal effects. The plant is attributed with antibacterial, anticancer, antidiabetic, antifungal, anti-inflammatory, antitubercular, antiviral, blood-purifying, gastroprotective, and immunomodulatory properties. Crude plant material as such, its extracts, oil, and purified principles are used alone or incorporated in various herbal formulations, ointments, soaps, creams, lotions, toothpastes, face washes, and other cosmetic preparations. These are used for the treatment of acne, pimples, boils, scabies, dermatitis, frunculosis, herpes, bleeding gums, pyorrhea, wounds, burns, malaria, dengue, and ulcers and for improving complexion. All parts of the tree, seed oil, and purified compounds are used for the treatment of cancer. Active principles are essential oil, nimbin, nimbidin, steroids, terpenoids, and sulfur. Many of the claimed pharmacological properties and therapeutic uses have been scientifically evaluated and confirmed (Bandhopadhyay et al. 2002; Parida et al. 2002; Gupta et al. 2006b).

    A recent review (Paul et al. 2011) on anticancer biology of this plant termed it as a "wonder drug" and nature's drugstore. Over 60 different types of chemical principles have been isolated from this tree. The available preclinical data largely pertain to preventive, protective, tumor-suppressive, immunomodulatory, and adaptogenic effects against various types of cancer and their molecular mechanisms. Reports on toxicity-related problems are inadequate: the authors recommended multicenter clinical trials to explore its anticancer potential and safety aspects.

    Reports for certain toxic effects of crude plant material, extracts, and Nimbin or Nimbidin are also available. These include allergenic effects during pollination season (rhinitis, asthma, dermal reactions), reprotoxicity, and biochemical or enzymatic changes in liver and kidneys of experimental animals. Oral LD_{50} value for oil in rats and rabbits were 14 mL kg^{-1} and 24 mL kg^{-1}, respectively. The animals showed respiratory distress, cyanosis, nasal secretions, and clonic convulsions before death with histopathological changes in the lungs and CNS on autopsy. Several cases of neem oil poisoning have also been reported in India. In a study at the Institute of Child Health and Children, Madras (1975–1980), ingestion of locally available neem oil (25–60 mL single dose), 12 children were hospitalized with convulsions and altered sensorium. Ten children died within 24 hours of oil administration (intended to control respiratory difficulty and intractable cough) (Sundaravalli et al. 1982; Gupta and Tandon 2004; Ashafa et al. 2012).

    Please also see examples 36 and 41 under Netherlands (Boeke et al. 2004) and South Africa (Ashafa et al. 2012), respectively, in Table 3.1 in Chapter 3.

15. *Berberis aristata* DC. (Daruharidra, Barberry). Stems and roots are claimed to possess analgesic, antidepressant, anti-inflammatory, antimicrobial, antipyretic, and antitrachoma properties (Chopra et al. 1980; Gupta 1989; Mishra 2004; WHO 2011). Some of these claims have been scientifically evaluated and confirmed. Berberine (active principle), an isoquinoline alkaloid, from this and some other plants, has a long history of medicinal use for treating varied disorders in India and China. These include diarrhea, dysentery, diabetes, neurodegenerative and neuropsychiatric disorders, eye diseases, for example, conjunctivitis, trachoma, and so on. Antidepressant activity was demonstrated in various behavioral paradigms of despair. The mechanisms involved included modulation of (1) Biogenic amines, (2) L-Arginine–nitric oxide–cyclic guanosine mono-phosphate receptor pathways, and/or (3) Sigma receptor pathway (Kulkarni and Dhir 2008a). Antidiarrheal effects are elicited through its action on *Escherichia coli* and heat-stable endotoxins causing gastric infection in Asia.

    Clinical studies have shown efficacy in cases of trachoma; berberine (0.1%–0.2% eye drops or ointment) cured 83.33% of patients. Only 50% of the patients became microbiologically negative, but most of the symptoms disappeared within three months of topical use. While no adverse effects or drug reactions were observed in adults following combined treatment with Neomycin or Sulphacetamide, children complained of burning sensation or irritation in eyes. The author stated that *Daruharidra* decoction is a useful remedy for treating conjunctivitis and chronic ophthalmic infection with no serious side effects (Gupta 1989). Some precautionary measures for safety have, however, been advised in a WHO-SEARO monograph (WHO 2011). These include (1) avoiding use of cosmetics, (2) removing contact lens, (3) hygienic measures, and (4) stoppage of drug if no relief is observed within three days of treatment and/or aggravation of symptoms. Though some decrease in locomotor activity and hypothermia were noted in experimental animals, many clinical reports indicate the drug to be fairly safe as no side effects were observed up to a dose level of 2 g per day.
16. *Boerhavia diffusa* L. (Punarnava). Whole plant and roots are used for anti-arthritic, anti-inflammatory, diuretic, and immunomodulatory effects. It attenuates myelosuppresion, inhibits apoptosis, and is reported to be useful in Alzheimer's disease and edema. The principal effects of the plant are its diuretic action with utility in edema. The activity principles are (1) alkaloid *Punarnavin* and (2) potassium (Seth and Sethy 1970; Chopra et al. 1980; Vohora and Wani 1987). The alkaloid *is* not very toxic; no untoward effects were observed in experimental animals, even with large doses. Clinical studies revealed beneficial effects in cases of edema and ascites. The drug elicited its optimal effect with the dropsic condition associated with healthy kidney. Seth and Sethy (1970) reviewed available data and stated that while no adverse effects were observed in these studies, there were several lacunae (an inadequate number of cases, controls, lack of electrolyte studies in urine, inconsistency of results when edema or ascites was due to cirrhosis, cardiac disease, or Kala-azar, etc.) for valid conclusions.

They recommended carefully controlled clinical trials to establish the efficacy and ensure the safety of *Punarnava* as a diuretic drug.

17. *Butea monosperma* (Lam.) Kuntz. Synonym: *B. frondosa* Koen. ex Roxb. (Palash, Dhak, Flame of the Forest). Flowers and seeds are attributed with anthelmintic, antifertility, antiestrogenic, parasiticidal, purgative, and tonic properties. Seeds are used to expel roundworms and hookworms from the stomach (Khanna and Chaudhury 1968; Chaudhury and Vohora 1970a; Chopra et al. 1980).

    The seeds may cause abdominal pain and giddiness (Saini 2004). Acute toxicity studies in doses ranging between 0.17 and 17 g kg^{-1} revealed mortality up to 30% in rodents within 72 hours. Chronic toxicity studies in rats, rabbits, and dogs exhibited progressive loss of weight, anemia, and congestive changes in the liver and spleen on postmortem examination. Suspension of seeds has been shown to exhibit teratogenic effects in rats. While no side effects were observed at recommended doses in humans, higher doses may cause nausea, vomiting, colic, and congestive changes in the liver, lungs, and spleen. Intake should be avoided in women who are desirous to conceive and pregnant women. However, its intake by the nursing mothers is quite safe for them and their babies (WHO 2011).

18. *Calotropis gigantea* L. Synonyms: *C. procera* L. (Aak, Madar, Milkweed). Leaves, stems, flowers, latex, and roots are considered poisonous but are also used for varied medicinal effects in the traditional system of medicine. These include asthma, gastrointestinal problems, diarrhea, helminthiasis, insecticidal, molluscidal effects, migraine, and cancer. Besides experimental studies, some clinical reports are also available. Capsules containing leaves from this plant, when taken for three consecutive days before sunrise on an empty stomach, exhibited a 100% cure rate in 50 patients with migraines. Surprisingly, the effects were negligible when the drug was taken during daytime (after sunrise). Other clinical reports showed the beneficial effects of its flowers in asthma and its dried root powder in gastric irritation, abrasions of intestinal mucosa, flatulence, dyspepsia, diarrhea, and dysentery, and in cases of various types of cancer (mammary carcinoma, osteosarcoma, cancerous growth in cheeks, adenosarcoma of ovaries, etc.). Best results were observed by the use of crude extracts in adenosarcoma of ovaries (Gupta et al. 2006a; Gupta and Sharma 2007).

    The juice of its plant has also been reported to elicit some adverse effects, for example, abortion and severe diarrhea. It is cattle poison, used to kill mad dogs and for suicidal purposes (Saini 2004). Cases of allergic contact dermatitis and poisoning with ocular lesions, ranging from superficial injuries and eyelid edema to severe eye damage (corneal ulcers, perforation, and loss of vision), have also been reported (Gupta and Sharma 2007).

19. *Camellia sinensis* L. (Chai, Tea). It is a very popular commonly used hot beverage by the masses. It is attributed with varied medicinal properties and therapeutic applications. These include antiallergic, anticancer, anticataract, anti-inflammatory, antimicrobial, antioxidant, antiulcer, refreshing, and stimulant effects. Caffeine is the active principle; other chemical

constituents are ascorbic acid, catechin, carotene, lipids, fatty acids, theoflavin, and so on. Effects of drinking tea, coffee, and caffeine were studied in human volunteers (18–22 years old) who were subjected to a constant work load of 610 kg M/min leading to fatigue. The work performance showed a marked enhancement in the following: (1) a prolongation of working period, (2) a shortening of the recovery period, (3) a lowering of blood lactate with fatigue, (4) a reduction in the α-index, and (5) a less synchronized pattern of electroencephalograph (EEG) in the treated group (Mitra et al. 1967; Koley et al. 1973, 1977). Beneficial effects were noted in periodontal disorders following the incorporation of tea dust in toothpastes and mouthwashes (Patel and Bhatt 1988).

Paradoxically, tea, which is attributed with anticancer properties, has also been shown to cause chromosomal damage and carcinogenic actions in mice and rats. Total aqueous extract (12 mg) and tannin-free extract (8 mg) given subcutaneously for 28 weeks produced tumors in 66% of treated rats (Kapadia et al. 1976). Mice exposed for 5, 10, and 15 days to caffeine and/or choline increased the frequency of sister-chromatid exchanges (SCEs) in bone marrow cells. Combined use showed additive effects (Panigrahi and Rao 1983). Dose- and duration-dependent potentiation were observed in fenfluramine-induced chromosomal aberration inhibition of the repair process in mice (Sen et al. 1994).

20. *Cannabis indica* L. Synonyms: *C. sativa* L. (Bhang, Ganja, Charas, Indian Hemp, Marijuana). Oral intake and smoking of leaves, flowers, hemp dust, and purified principles from this plant cause intense CNS depressant effects. The main chemical constituents are cannabinol and cannabindiol. The drug is widely abused by masses particularly in rural areas, Holi (festival of color) season in India, and by religious groups (*Sadhus*). Its pharmacological properties (analgesic, anticonvulsant, hypnotic, narcotic, sedative) and adverse or toxic effects (addiction, allergy, catalepsy, chromosomal damage, dependence, ototoxicity, psychosis, reprotoxicity, teratogenicity, and tolerance) have been extensively investigated in experimental animals and human studies (Gupta et al. 1974; Jha et al. 1975; Vijaylakshmi and Singh 1975; Shankar et al. 1978; Singh et al. 1982; Malik et al. 1983; Gupta and Gupta 1994; Sharma et al. 2000; Gupta and Sharma 2007).
21. *Capsicum annuum* L. Synonyms: C. *frutescence* L. (Lal Mirchi, Red Chili). Fruits are extensively used in spicy Indian food, curries, and pickles and eaten raw (green variety). These are attributed with varied medicinal properties, for example, anthelmintic, antimicrobial, antiviral, appetizing, antioxidant, antihyperlipidemic, cholinergic, and insecticidal actions. Some reports of their utility in diabetes mellitus and hepatoprotective and radioprotective effects are also available. The chemical constituents capsicum (amide) and paprika (oleoresin) are considered to be its active principles (Gupta and Sharma 2007). Several adverse or toxic reactions have been reported. Chilies are very pungent (due to their amide content) and produce lachrymatory (tears), pro-inflammatory (on gastrointestinal and respiratory systems), mutagenic, and carcinogenic effects (Sirsat and Khanolkar 1960; Nagabhusan and Bhide 1985; Agarwal and Bhide 1987, 1988; Bhide et al.

1993; Chitra et al. 1994; Saroja et al. 1995; Nalini et al. 1998). Excessive use of chilies may cause (1) diarrhea, a burning sensation or pain during defecation, (2) hyperacidity, (3) gastric ulcers, (4) bleeding piles, (5) hepatitis, (6) cystitis, nephritis, and (7) induction of premature labor in pregnant women leading to miscarriage (Gupta and Sharma 2007).

22. *Carathamus tinctorius* L. (Kusum, Safflower). It activates the digestive enzyme amylase, has nutritive value, and is claimed to possess anthelmintic, antimicrobial, cardioprotective, and insecticidal properties. Seeds infected with *Fucarium oxyspora* have been reported to contain epoxytrichloranthracene. When the oil from such seeds (0.1, 0.2, and 0.4 mg) was applied on the shaved skin of albino rats, it resulted in dermotoxic effects: reddish wheals on skin (Gupta and Sharma 2007).
23. *Carica papaya* L. (Papita, Papaya). Fruits and principles from this plant are attributed with antibacterial, antifungal, anticancer, anticoagulant, antifertility, antimalarial, hypolipidemic, and hepatoprotective properties. These may also cause adverse or toxic effects, for example, abortion, allergy (to skin and respiratory system up on exposure to pollen), genotoxic and treatogenic effects, bronchial asthma, and damage to respiratory and reproductive systems. The chemical constituents are Benzyl-thiocyanate, β-carotene, Carpesmine, Carpine, and so on. (Shivpuri and Dua 1963a,b; Bhagat et al. 1993; Gupta and Gupta 1994; Chinnoy et al. 1997; Dalsaniya et al. 1999; Gupta and Sharma 2007).
24. *Cassia* spp., *Cassia absus* L. Synonyms: *C. coccinia* Wall. (Chaksu), *C. carandus* L. (Karonda, Christ's Thorn), *C. fistula* (Amaltas, Indian Labernum), *C. occidentalis* L. (Badi Karondi, Coffee Senna), *C. siamea* Lamk. (Ironwood Tree), and so on, contain Carisine, Chaksine, Emodin, Narigine, and Sennosides. These are attributed with varied medicinal properties and uses in Indian systems of medicine. These include analgesic, anthelmintic, anticancer, anti-inflammatory, antimalarial, antimicrobial, antioxidant, antitussive, cardiotonic, hypoglycemic, hypotensive, insecticidal, and nutritive actions. Some toxic or adverse effects have also been reported in experimental animals and human subjects: (1) seasonal allergy (asthma, fever, rhinitis, skin reactions) in various parts of the country, (2) gastrointestinal problems (diarrhea, vomiting), (3) dermatitis (observed and positive patch tests in many cases), and (4) exhaustion and death reported with ethanolic extract of roots in cats. Some of the effects were attributed to histamine release following exposure (Saha and Kalyansundaram 1962; Bajaj et al. 1982; Chaubul and Gadve 1984; Bhagat et al. 1993; Gupta and Gupta 1994; Sateesh et al. 1994; Dalsaniya et al. 1999; Gupta and Sharma 2007).
25. *Celastrus paniculatus* Willd. (Malkagni, Staff Tree). The husk and seeds are used. These contain palmitic acid, stearic acid, and fixed oils. The chemical constituents present in the latter include oleonolic, linoleic and linolinic acids; β-sitosterols; sesqueterpenes; polychols, and so on. The plant is attributed with anticonvulsant, anti-inflammatory, antimicrobial, hypolipidemic, and insecticidal properties and claimed to be useful in atherosclerosis. This drug has also been reported to produce antispermatogenic, hepatotoxic, and

nephrotoxic effects. Semipolar and polar compounds, present in chloroform and methanol extracts (0.6 mL diluted with equal quantity of edible oil, given intraperitoneally [i.p.] once a week for a month), caused fatty degeneration in the liver and proximal tubular damage of kidneys in treated rats. This was accompanied by biochemical changes (increase in serum serum glutamic oxaloacetic transaminase (SGOT), serum glutamic pyruvate transaminase (SGPT), and alkaline phosphatase) (Bidwai et al. 1990a,b; Gupta and Sharma 2007).

26. *Cenchrus* spp. *C. biflorus* Roxb. (Bhront), *C. cillaris* L. Synonyms: *Pennisetum cenchroides* LC, Rich. (Anjan, Buffel Grass), *C. setigerus* Vahl. (Anjan, Birwood Grass). The leaves and buds contain Ascorbic acid, Alkaloids, Flavonoids, Nutritive proteins, Saponins, Tannins, and so forth, and are attributed with antibacterial and antifungal properties. These have been confirmed by scientific studies. While LD_{50} of the ethanol extracts was found to be more than 1 g kg^{-1} i.p. in rodents and showed no untoward effects in experimental animals (Gupta and Sharma 2007), some allergenic effects were observed in humans. Out of 75 patients tested for rhinitis and asthma, pollens from this plant were reported to be the cause of allergy in 6.5% of patients in Kolhapur during a monsoon (Sateesh et al. 1994). Further intradermal tests to study the airborne grass pollens for antigenic or allergenic activity revealed marked positive skin reactions (22%) and enhanced specific Ig E levels (Sridhara et al. 1995). Clinically significant positive reactions in skin tests (18.2% and 54%, in scratch and intracutaneous tests, respectively) were observed with *Cenhrus* grass in 100 cases of rhinitis and asthma that were reported in Delhi (Shivpuri and Dua 1963b).
27. *Centella asiatica* Urban. Synonym: *Hydrocotyle asiatica* L. (Brahmi, Manduki). Whole plant and roots are attributed with anabolic, antiepileptic, antitumor, nootropic, tranquilizing, and wound-healing properties. The plant is highly reputed for its brain tonic properties in Ayurveda; incorporated in many memory-improving formulations that are sold in India; and used for the treatment of Alzheimer's disease, mental retardation, and convulsive disorders (Vohora 1989; Vohora and Mishra 2004). The plant is used by Naga tribals for the treatment of diarrhea and cholera (Jamir 1989).

    Cytotoxic and antitumor properties were reported in certain *Taxa* of the family *Umbelliferae*, especially *C. asiatica* L. The effects observed with crude extracts or partially purified fractions *in vitro* and *in vivo* can be used to treat cancer as no discernible toxic effects were observed in normal human lymphocytes (Babu et al. 1995). Radiation is known to cause behavioral perturbation like conditioned taste aversion (CTA); performance and learning protection against CTA, reported with this plant, may also be of immense value by its concurrent use in patients who are undergoing radiotherapy (Shobi and Goel 2001).

    Please also see example 30 under Italy in Table 3.1 in Chapter 3 (Cravotto et al. 2010).
28. *Cerebra manghas* L. Synonym: *C. odollam* Gaertn. (Pili Kirbir, Dog Bane). Seeds and kernels contain cerebrin, glycosides, fatty acids, saponins, and

β-stiosterol. Cerebrin is considered the active compound. While the plant showed some cardiotonic and anticonvulsive effects in experimental animals, it has also been reported to cause spasmogenic and cardiotoxic actions (Gupta and Sharma 2007). Four cases of suicidal poisoning by consumption of seed kernels were reported in *Kerala*. The symptoms included nausea, vomiting, and electrocardiographic abnormalities before death. The effects, attributed to cerebrin in fruits, were comparable to those that are usually caused by digitalis (Kino and Pai 1965). Another two cases from *Kerala* were reported with similar symptoms within 8 hours of ingestion of kernels (Kuntali et al. 1970). Six patients, who consumed this drug and showed cardiac abnormalities, were studied for neurological complications. Trivial effects (hypotonia, hyperreflexia, and nonspecific slowing of EEG) were observed suggesting depressant action on peripheral nerves or at spinal synapses (Iyer and Narendranath 1975).

29. *Cerebrum noctutnum* L. (Rajnigandha, Poison Berry). Flowers and leaves contain essential oil, Saponins, Aglycones (Tigogenin and Yuccagenin), Steroids, and Terpenoids and are attributed with antimicrobial, cardiotonic, hemolytic, and nutritive properties (Gupta and Sharma 2007). Some allergenic or toxic effects have also been reported, for example, contact dermatitis (Pascricha and Kanwar 1978).
30. *Cinnamomum zeylanicum* L. Synonym: *Cassia cinnamon* L. (Dalchini, Cinnamon). Bark, leaves, and cinnamon oil are attributed with astringent, digestive, and antiflatulent properties and used for the treatment of varied ailments. The principal therapeutic uses are in enterotoxic diarrhea and respiratory problems, but the drug has also been used for the treatment of gum disease, nervous tension, and infections (especially vaginal infection) and for improving complexion and memory. The plant is commonly used as a flavoring agent in curries, cakes, and sweets and as a food (Chopra et al. 1980; Hussain 1989; Bakhru 2001; Mitra and Rangesh 2004a,b).

    An online cinnamon challenge dared people to swallow a spoonful of cinnamon powder without washing it down with water. It resulted in serious symptoms, for example, dry mouth, intense pain, choking cough with sputtering of terracotta smoke, suffocation (which may be life threatening), liver disorders, and so on. The challenge, though not new, has alarmingly revived since 2001 with several participants (including celebrities) taking it. More than 30,000 videos are on YouTube showing the suffering participants. This has prompted the U.S. health authorities to warn against such attempts (EU Food Law 2006). Toxic effects have also been reported on renal function (damage to kidneys and urinary bladder), headache, and abortion. Some precautionary measures have been advocated by Indian physicians who advise against its use by persons with hot constitution (Bakhru 2001). Coumarines are responsible for its toxic effects (Abraham et al. 2010).

    Please also see example 18 under Germany in Table 3.1 in Chapter 3 (Abraham et al. 2010).
31. *Cissus pareira* L. (Patha, Laghupatha). Root, bark, and leaves are used as a muscle relaxant and psychostimulant. Utility for endotracheal intubation

during surgery has also been reported. Hayatin and its derivatives are active principles of the plant (Chopra et al. 1980; Amresh et al. 2008). While no adverse or toxic effects were discernible in acute and sub-acute toxicity, studies were recommended by researchers from the National Botanical Research Institute, Lucknow, to validate (or refute) the safety claims for this medicinal plant (Amresh et al. 2008).

32. *Cissus quadrangularis* Wall. Synonyms: *Vitis quadrangularis* Wall., *Heliotropium indicum* L. (Asthiandhana, Hadzod, Har Singar). Roots, stem, and leaves are used for androgenic effects and healing of wounds and fractured bones. Such effects were demonstrated by detailed experimental and clinical studies that were carried out at S.S. Hospital, Banaras Hindu University, Varanasi. The investigations used a range of techniques including biochemical, isotopic tracer studies (with radioactive isotopes of Ca, P, S, Sr, and Proline), measurement of tensile strength, microangiography, and tissue culture. Efficacy was shown by both oral route and local application of a paste that was made from the plant. Studies at King George Medical College, Lucknow, showed earlier radiological healing (approximately four weeks) and tensile strength judged by biting force in dental cases. Phytogenic steroid (isolated in crystalline form) was considered the active fraction. Presence of calcium (as CaC_2O_4) and high vitamin C content contributed to its healing effects (Pramod and Udupa 1970; Vohora 1979; Saxena et al. 1989).

    No serious adverse effects were observed in most patients. Application of paste caused local itching sensation in the same patients. It subsided within one to two days without any medical aid (Pramod and Udupa 1970).

33. *Commiphora mukul* Engl. Synonym: *Balsamodron mukul* Hook. (Guggul). It is an exhaustively studied drug, used as a crude extract, purified fractions or compounds, a single plant, and an ingredient of compound formulations. It is attributed with antiarthritic, anti-inflammatory, hypocholesterolemic, and hypolipidemic effects. Experimental and clinical studies revealed efficacy in arthritis, inflammatory conditions, lipid disorders, thrombosis, heart disease, and obesity. Steroidal fraction, isolated from petroleum ether extract, showed potent antiarthritic activity that is comparable to hydrocortisone and more potent versus phenyl butazone. Gum resin, steroid fraction, and unidentified crystalline compounds were considered responsible for its therapeutic value (Tripathi 1970; Vohora 1979; Chopra et al. 1980; CHEMEXCIL 1992). During the development of this herbal drug as a hypolipidemic agent, the crude extract caused minor side effects (diarrhea, menstrual disturbances, skin rashes, etc.). Purification by techniques described in Ayurvedic texts lead to disappearance of such ill effects. Boiling of drug in water and *Triphala powder* was used for this purpose. The powder is a mixture of three fruits (*Emblica officinalis, Terminalia belerica,* and *Terminalia chebula*) (Sharma et al. 2009).

    Please also see examples 29 under Italy in Table 3.1 in Chapter 3 (Grieco et al. 2009).

34. *Coriandrum sativum* L. (Dhania, Coriander). Whole plant, seeds, and volatile oil are used medicinally. It is a flavoring agent that is commonly used in curries and soups and for preparing sauces or chutneys. The plant is rich in vitamin A, B_1, B_2, and C and iron and catarrh properties. No toxic effects were reported. It is, however, advisable to avoid its excessive use by patients of asthma and bronchitis. These diseases may be aggravated by such intake (Bakhru 2001).
35. *Crocus sativus* L. (Kesar, Zafran, Saffron). Flowers, bulbs, and stigmas are precious and used medicinally for varied purposes. In addition to its use as a coloring or flavoring agent in food, sweets, and pharmaceutical preparations, it is used for its aphrodisiac and antispasmodic effects. It possesses stomachic and tonic properties, promotes libido, and is useful in spasmodic disorders and sexual weakness. Excessive use may, however, elicit narcotic effects. Its use should be avoided in children and pregnant women (may cause abortion) (Chopra et al. 1980; WHO 2011).
36. *Curcuma longa* L. (Haldi, Turmeric). Rhizomes and curcumin are highly reputed for antiarthritic and anti-inflammatory effects. Considerable experimental and clinical data are available to support their efficacy in arthritis and inflammatory diseases. Other pharmacological properties attributed to this plant include anthelmintic, antibacterial, antifungal, antioxidant, hypoglycemic, hypolipidemic, antifertility, anticancer, hepatoprotective, and radioprotective effects and utility in treating bronchial asthma, hypothyroidism, and tumors (Bhatia et al. 1964; Srimal and Dhawan 1973; Garg 1974; Banerjee and Nigam 1978a,b; Jain et al. 1990; Ali et al. 1995; Anto et al. 1996a,b, 1998; Godkar et al. 1996; Deshpande et al. 1998; Choudhary et al. 1999; Joseph et al. 1999; Bhagat and Purohit 2001; Balasubramanyam et al. 2004; Jagetia and Rajnikant 2004; Sharma et al. 2009). Some studies have probed the enzymatic or molecular mechanisms that are involved.

    Reports of some toxic effects are also available in experimental animals and allergic skin rashes and fever in human subjects (Bhawani Shankar et al. 1980; Jain et al. 1987; Ali et al. 1995; Joshi et al. 2003). Deterioration of quality during storage has also been probed (Srimal and Dhawan 1973). A WHO report (WHO 2011) states that though turmeric is being regularly used in Indian food and is considered in safe doses, exceeding recommended doses may cause gastrointestinal problems. It is advisable to avoid the intake of this herbal drug on an empty stomach and by patients with blockage of bile duct, gastric ulcers, and coagulation disorders. Concurrent use with aspirin and warfarin may cause bleeding in susceptible persons. When no benefit is observed within two to three days, the drug should be stopped and medical advice sought. Two recent reports present contradictory finding. Research from Cadila Pharmaceuticals Ltd., Dholka, carried out a safety assessment of a solid lipid curcumin particle preparation. The no-observed-adverse-effect level for this standardized preparation was found to be 720 mg kg^{-1} body weight per day (the highest dose that was tested in rats) (Dadhaniya et al. 2011). Scientists from the Department of Biotechnology, Manipal University, Manipal, investigated 200 chemical compounds from turmeric for toxicity prediction: 184 compounds were predicted toxigenic, 135 mutagenic, 153 carcinogenic, and 64 hepatotoxic. To cross-validate their results, these workers selected curcumin (the most popular compound) and found that it and its derivatives may cause hepatotoxicity in a dose-dependent manner (Balaji and Champaran 2010).

37. *Datura metal* L. Synonym: D. *fastuosa* L. (Dhatura, Thorn Apple). Leaves, seeds, and roots are used for varied medicinal effects. The principal use is for the treatment of filariasis and helminthiasis. Other pharmacological properties, attributed to this drug, include analgesic, antispasmodic, antiviral, aphrodisiac, and narcotic or sedative actions with utility in skin diseases, lice infestation, ulcers, earache, and sexual weakness (Hussain 1989; Hameed and Vohora 2001; Gupta 2002; Gupta et al. 2008; WHO 2011). Seed powder and aqueous extracts were shown to produce hypoglycemic (normal and alloxan diabetic) and hypocholesterolemic effects in rats (Gupta 2002). Withafustuosin D and E (isolated from the leaves) exhibited antistress anxiolytic and antiulcer effects in rats (Maiti et al. 1997; Manickam et al. 1997).

    It is a poisonous plant and should be used with care. All parts of the plant are toxic; seeds are the most toxic. Ingestion of 100 seeds may be fatal; intake of 4 g of plant material may be lethal to children. While applying or removing leaf juice or paste of seeds, care should be taken to ensure that it does not get into the eyes, mouth, ears, nose, and other natural openings of the body. Mixing this plant with other drugs having similar actions is not desirable. It may lead to serious drug interactions. Lice should first be removed manually using a comb before using this drug. In case only a few lice are found following two to three applications, there is no need to continue the treatment. Hyocyamine, hyosine, and atropine are responsible for its medicinal and toxic effects. Physostigmine (0.5–2 mg in adults and 0.02 mg kg^{-1} by slow intravenous injection) may be used as an antidote to *Datura* poisoning (Rates 2001; Hameed and Vohora 2001; WHO 2011). Chronic treatment with ethanolic extract of seeds increased rat brain lipid peroxidase and catalase activity. It also impaired the activities of aldolase and glucose-6-phosphate dehydrogenase and decreased nucleic acid metabolism (Hasan and Kushwaha 1987).

    Please also see example 10.4.6 under Brazil in Table 3.1 in Chapter 3 (Rates 2001).

38. *Datura stramonium* L. (Datura) is a sedative and narcotic plant that is used for varied medicinal effects and as an intoxicant. Large doses may be fatal (Saini 2004).

39. *Eletteria cardamomum* L. (Chotti Elaichi, Cardamomum). Seeds, bark, and essential oil are attributed with analgesic, antiemetic, antispasmodic, appetizing, carminative, digestive, antimicrobial, and anti-inflammatory and stimulant properties. It is commonly used as a flavoring agent in food items, tea, sweets, and "pan" and used to treat digestive (nausea, vomiting, indigestion) and inflammatory problems and in bronchial asthma (Chopra et al. 1980; Lodha and Kabra 2004; WHO 2011).

Traditionally, cardamom seeds are considered safe owing to their long use in food and beverages. While no adverse effects have been observed, some reports indicate that the seeds can trigger gallstone colic and so are not recommended in patients with gallstones. Medication with this herbal drug should be restricted to mild nausea and vomiting from gastrointestinal causes and pregnancy. Severe cases may have other underlying causes (including kidney failure, stomach cancer, brain tumor, psychogenic vomiting, etc.) for which proper medicinal advice should be sought (WHO 2011).

40. *Embellia officinalis* L. Synonym: *Phyllanthus emblica* L. (Amla, Amlaki, Indian Gooseberry). Fruits are attributed with varied medicinal properties and therapeutic uses. These include anabolic, anticarcinogenic, antioxidant, cooling, and immunostimulant actions. It is highly reputed for boosting general resistance of the body to diverse infections and diseases. It is a rich source of vitamin C for use in nutritional deficiencies, pregnancy, and chronic ailments. The active principles are ascorbic acid, phyllemblin, and terpenoids. The fruits are used as such and as a prime ingredient of a popular Ayurvedic Rasayana formulation: *Chyvanprash*. The latter is manufactured and marketed by many reputed pharmaceutical companies on the Indian subcontinent. It is promoted to be a general tonic and prevent or treat varied diseases including gastrointestinal problems, acidic gastritis, duodenal ulcers, chronic infections, respiratory problems, and general debility (Vohora 1979; Chopra et al. 1980; Vohora and Wani 1987; Panda et al. 2002; Mungatiwar and Phadke 2004; WHO 2011).

    Amlaki powder is generally a safe medicine. No toxic or adverse effects have been reported even with continuous use. Ethanolic extract produced no cellular toxicity. Use by a large population for varied indications points to its safety in all age groups, including pregnant and nursing women, infants, and children. A WHO monograph (2011) recommended some precautionary measures: (1) Because of its coolant properties, individuals intolerant to cold should take it along with ginger powder with honey in warm water. (2) It is slightly bitter or sour in taste. Mixing with sugar or honey makes ingestion of the powder easier.

41. *Eugenia jambolana* Lam. Synonym: *Syzgium cumini* L. (Jamun, Black Plum). Fruits, seeds, and flowers are used. The principal therapeutic value for this herbal drug is for the treatment of diabetes mellitus. Besides hypoglycemic and antihyperglycemic effects, it has also been reported to have antiallergic, antifertility, antimicrobial, and antioxidative properties. Experimental studies confirmed its principal activity against alloxan- or streptozotocin-induced hyperglycemia in rats and rabbits. Clinical studies revealed marked lowering of blood sugar levels in non-insulin-dependent diabetes mellitus (NIDDM) and associated cataract. Highly active compounds have been isolated from its fruits and pulp. Increase in Cathepsin B activity and insulin secretion were proposed as the mechanisms that were involved. The antifertility effects were attributed to the oleoneolic acid that is present in the drug (Brahamchari and Augustu 1961; Mukherjee et al. 1963, 2004; Chaudhury and Vohora 1970a,b; Vohora et al. 1973; Vohora

1979; Chopra et al. 1980; Bansal et al. 1981; Rathi et al. 2002; Kar et al. 2003; Mishra and Adra 2004; Vohora and Vohora 2005; Bagiga et al. 2011).

42. *Ferula asafetida* (Heeng, Devil's Dung, Asafetida). Dried latex and resinous gum are used as a condiment, relished in India and Iran. It is commonly used as a flavoring agent in curries, meatballs, and pickles. The plant is also attributed with antiflatulent, antispasmodic, aphrodisiac, and nervine-stimulant properties. It removes gas and associated gastric pain. Excessive use may cause toxicity: severe vomiting and dehydration. It should not be given to infants and children (Bakhru 2001).
43. *Ficus bengalensis* L. (Bargad, Vat Vraksha, Banyan Tree). The bark is attributed with hypoglycemic or antihyperglycemic properties with utility in diabetes mellitus. This has been validated by some scientific studies. Investigations in experimental animals (mice, rats, rabbits, and dogs) confirmed these effects. A linear dose–effect relationship was observed with the ethanolic extract of bark. No activity was observed in depancreatized animals and when detannated extracts were used. Lack of activity with tannin-free extracts indicates that one of the mechanisms involved was inhibition of glucose absorption from the gastrointestinal tract (GIT). This raises the question whether the absorption of other essential nutrients was also inhibited. No activity in depancreatized animals suggests that the mechanism of action was similar to that of Tolbutamide. Aqueous extract showed no effect on glucose uptake in rat diaphragm. Inhibitory action was noted against insulin-degenerative process. Further stimulant activity was observed on insulin secretion from the islets of Langerhans in the pancreas. The active compounds isolated from the plant include β-sitosterol; flavonoidal compounds A, B, and C; glycosides in fraction B, lecucoanthocyanin-3-0-delta galactosyl glycoside, quercetin-3-galactoside, and pellarigonidin. The last compound was found to be more effective versus others principles that are isolated from the plant (Chaudhary et al. 1961; Mishra and Adra 2004). While no serious adverse or toxic effects have been reported, Mishra and Adra (2004) recommended at least the following studies on traditional herbal drugs or formulations that are used for the treatment of diabetes mellitus: (1) LD_{50}, (2) subchronic toxicity for 120 days in normal and diabetic animals, (3) life-term animal studies to probe effect on survival time, effect on pancreas (protective or toxic), and (4) clinical trials under controlled conditions.
44. *Glycyrrhiza glabra* L. (Mulethi, Yarti Madhu, Licorice). Roots or tubers and glycyrrhizin are used for medicinal purposes. Its principal utility is for the treatment of respiratory ailments (expectorant and antitussive action), but this herbal drug is also attributed with various properties: antiarthritic anti-inflammatory, antimicrobial, (bacteria and viruses), antimutagenic, febrifuge, and immunomodulatory actions. It has been used for the treatment of cough and cold, gastrointestinal problems (diarrhea, hyperacidity, gastric or peptic ulcers), gynecological problems (menopausal syndrome), tumors, Alzheimer's disease, and HIV. Glycyrrhizin inhibits inflammation and prostaglandin synthesis. The drug is reported to enhance

availability of estrogen after menopause (Vohora 1979; Chopra et al. 1980; Naqvi et al. 1991; Mungantiwar and Phadke 2004; Kaur et al. 2009, 2012; Chandrasekaran et al. 2011; Shah et al. 2011; WHO 2011).

Though licorice roots are widely used in cough syrups for soothing irritation both in the throat and stomach, prolonged use in large doses has been reported to cause hypertension and edema (with puffiness in the body) in some patients. This undesirable effect has been attributed to the main saponin principle in the roots, glycyrrhizin, its sapogenin, and glycyrrhitinic acid. It is advisable to avoid its use in pregnant women and in patients with hypertension and renal disease (Premila 1989; Bakhru 2001; Luyckx 2012). Licorice interacts with Spironolactone, which offsets the former's effects (Parmar 2005).

Please also see example 32 under Japan in Table 3.1 in Chapter 3 (Homma et al. 1993).

45. *Gymnema sylvestre* R. Br. Synonym: *Ascelepias germinata* Roxbin (Gurmor). Roots and leaves as such, and their crude extracts have been used extensively for the treatment of diabetes mellitus and obesity on the Indian subcontinent, Australia, Japan, and Vietnam. Other properties attributed to this herbal drug include antiallergic, antibacterial, and antiviral effects. Experimental studies in normal and against induced hyperglycemia (alloxan, streptozotacin, and pituitary extract) revealed marked action on fasting blood sugar (comparable to Tolbutamide) and favorable effect on glucose tolerance, glucose homeostasis, glycosuria inhibition, and longevity in experimental animals. Preventive action was also observed against alloxan-induced cataract in rats. Clinical studies showed moderate activity in normal subjects and diabetic patients with the aqueous extract in NIDDM and IDDM patients. Enhancement of endogenous insulin (secretion and release) and promotion of regeneration of islet-cells of pancreas were suggested. The active compounds are GS-4, gymnemic acid, gurmarin (a polypeptide of 85 amino acids), lupeol, stigmasterol, saponins, polysaccharides, and terpenoid sweetness inhibitors (Wahi and Chunekar 1964; Vohora 1979; Chopra et al. 1980; Bone 1996; Rathi et al. 2002; Kar et al. 2003; Porchezhian and Dobriyal 2003).

    Adequate safety data are not available on this herbal drug with wide usage. This aspect needs scientific attention. The plant extracts can temporarily reduce or abolish taste sensation of sweetness or bitterness. Some efforts have been made to mask its bitter taste in preparations containing this plant. A detailed review is available on its chemistry, pharmacology, and patients (Porchezhian and Dobriyal 2003).

46. *Holarrhena antidysentrica* Wall. (Kurchi, Kutaja, Tellycherry). Bark and seeds are attributed with antiamoebic, antidiarrheal, antidysenteric, and astringent properties. The principal therapeutic use of this herbal drug is in irritable bowel syndrome and gastrointestinal problems (including enterotoxic diarrhea, irritable colon, and *E. coli* infection). It has been used as a single drug and in combination with other herbal ingredients in Ayurvedic formulations. Experimental and clinical studies are available to support its

efficacy on dysentery (*Pravatika*) and irritable bowel syndrome (*Grahni*). A recent report described its use in cases of malaria. This plant contains alkaloids—conamine, conkurchine, conessimine, holarrhemine, and so on—which are responsible for its medicinal properties (Vohora 1979; Mitra et al. 2004a,b; Verma et al. 2011; WHO 2011).

While no side effects or toxicity have been mentioned in classical Ayurvedic texts and in some clinical studies with *Kutaja* powder at recommended doses (both as a single drug and in formulations), it may cause distention of the abdomen after using for a few days. A WHO monograph (2011) mentioned some constipatory action on prolonged use and hypertension in some patients. It is recommended to (1) discontinue the drug soon after the diarrhea is checked and (2) use with caution or under medical supervision in pregnant women and hypertensive patients. Ingestion by the nursing mother is, however, safe for the baby.

47. *Hypericum perforatum* L. (Basanti). The plant causes dermatological problems and blisters. Hair loss has been observed in animals (Saini 2004).
48. *Jatropha curcus* L. (Ban Rend). Intake of seeds causes diarrhea (Saini 2004).
49. *Lathyrus* spp., *L. aphacea* L. (Ankari), *L. sativa* (Latari). Ripe seeds are narcotic. Toxicity with whole plants causes lathyrism in man and animals (Saini 2004). Some popular pulses used as food in India have been reported to be adulterated with cheaper products causing such poisoning (Vohora 1974).
50. *Lepidium sativum* L. (Chamsur Halun). It is claimed to possess varied properties, but most of these claims have not been investigated. Vohora and Khan (1977) carried out studies on some Unani herbs for their antifertility effects and found no antiovulatory activity in this plant. Pharmacological investigations on the ethanolic extract of seeds, by the same group, revealed cardiac and skeletal muscle stimulant effects. The presence of a cardioactive substance was suggested, which was found to be unstable in solution and exhibited tachyphylaxis in anesthetized cats. This effect was possibly mediated through adrenergic mechanisms. No gross or behavioral toxic effects were discernible in doses up to 1 g kg^{-1} i.p. in mice (Vohora and Khan 1977). Seeds given with milk have been used to induce abortion (Saini 2004).
51. *Momordica charantia* L. (Karela, Bitter Gourd). Fruits are highly reputed for their hypoglycemic or antihyperglycemic actions for therapeutic use in diabetes mellitus. Many experimental and clinical studies are available to support its value in this disease. Investigations were carried out in normal and various models of experimental diabetes (alloxan, streptozotocin, anterior pituitary extract, and glucose feeding) using fresh fruit juice, crude extracts, and purified principles (charantin, karara, momordin, oleonolic acid-3-0-monodesmoside, and polypeptides). The purified principles revealed more potent activity versus tolbutamide. Clinical studies showed moderate activity in NIDDM patients. Pancreatic and extrapancreatic mechanisms have been proposed. These include (1) improved insulin utilization and (2) decreased insulin resistance. Synergistic effects were

reported with metformin and glibenclamide. Besides its antidiabetic properties, the plant has been attributed with favorable activity on lipid profile and thyroid function (Chaudhury and Vohora 1970b; Vohora et al. 1973; Vohora 1979; Chopra et al. 1980; Mishra 2004; Mishra and Adra 2004; Mukherji et al. 2004; Vohora and Vohora 2005).

LD_{50} of 50% ethanolic extract of fruit was found to be 68.1 mg kg^{-1} in mice. Detailed toxicity studies are not available. Large quantities of the extract induced lesions and affected testicular function in dogs. Abortifacient effects and death of pregnant rabbits (due to uterine bleeding) were reported. Adverse effects, observed in clinical studies, range from mild headache to serious reactions (e.g., hypoglycemic coma and convulsions) in children. This calls for caution and warrants further investigation and monitoring. Detrimental interactions may occur on concurrent use with modern antidiabetic medicines (Dixit et al. 1978; Raman and Lau 1996; Tongia et al. 2004; Gupta et al. 2005).

Please also see example 47 under the United Kingdom in Table 3.1 in Chapter 3 (Raman and Lau 1996).

52. *Mucana pruriens* L. (Kappikachu). Seeds and fruits are attributed with aphrodisiac, anti-inflammatory, hypocholesterolemic, and nervine tonic properties. This herbal drug is useful for treating sexual dysfunction and nervine disorders including Parkinson's disease. Experimental studies, using seed powder (75 mg kg^{-1} per day), showed marked enhancement of sexual activity in male rats. Different components of copulatory behavior (mount frequency and latency, intromission frequency and latency) were influenced by the drug. The plant contains e-DOPA, which is reported to arouse sexual desire in patients who are suffering from Parkinson's disease. Other notable chemical compounds present in the plant are alkaloids and β-sitosterol. The alkaloids have promoted sperm count and increased weights of testes and prostate glands. No citable clinical reports are available (Akbarsha et al. 2004; Mishra and Singh 2004; WHO 2011). Some adverse effects (diarrhea and weight loss) have been reported on prolonged use. These were attributed to the protein fraction. Judicious use at recommended doses is well tolerated. Discontinuation of the drug may be necessary in case of unchecked diarrhea (WHO 2011).

    Please also see example 10 under Brazil in Table 3.1 in Chapter 3 (Rates 2001).

53. *Myristica fragrans* (Nutmeg). Dried seeds, kernels, and oil are used as flavoring agents in food, liniments, perfumery, and hair lotions. The drug is attributed with carminative and antispasmodic properties. It should be taken in small doses; intake of even a teaspoon may induce toxic symptoms, for example, gastritis, vomiting, restlessness, giddiness, and hallucinations. Large dosage excites the motor cortex and may lead to epileptic convulsions and produce lesions in the liver (Bakhru 2001).
54. *Nardostachys jatamansi* DC. (Jatamansi). Roots and seeds are attributed with adaptogenic, antiarrhythmic, anticonvulsant, antihypertensive, and antioxidant properties. The drug has been reported to be useful in

Alzheimer's disease, bronchial asthma, as a protective agent against liver damage and cerebral ischemia, gynecological disorders (including post-menopausal problems), nervine disorders (e.g., Parkinson's disease), and skin diseases. The active constituents isolated from it are Jatamansone and Valaranone (Arora 1965; Vohora 1979; Gupta 1989; Ali et al. 2000; Hameed and Vohora 2001; Salim et al. 2003; Rao et al. 2005; Ahmad et al. 2006; Lyle et al. 2009; WHO 2011; Sharma and Singh 2012).

LD_{50} of the ethanolic extract of roots was found to be 569 ± 20.8 mg kg^{-1} i.p. in albino rats. No mortality was found following oral intake of the extract up to 1 g kg^{-1}. The extract affected the central and autonomic nervous system and behavior during the first 4 hours after oral and i.p. administration in a dose-dependent manner. At a dose of 500 mg kg^{-1} i.p., it caused loss of righting reflex. No consistent loss of body weight and other untoward effects were observed (Singh and Rastogi 1972; Rasheed et al. 2010).

55. *Papaver somniferum* L. (Afim, Opium, Poppy). Seeds and roots are used for medicinal purposes and as an intoxicant. The seeds are a source of fatty oil and have nutritive value for use in food, breads, sweets, confectionery, and curries. These are attributed with cooling, thirst quenching, and febrifuge properties and are useful in gastric irritation, inflammation, biliary problems, pain, and nonspecific fevers and for the control of vomiting in cholera. Roots are stimulant and tonic. Excessive use is harmful and causes intoxication. It should not be used in (1) infants or children, (2) pregnant women, (3) sensitive individuals, and (4) patients with renal disease (Bakhru 2001).
56. *Picrorhiza kurroa* Royle. ex Benth. (Kutki, Katuka). The roots, extracts, and active principle—Picrorhizin or Kutkin (an iridoid glycoside)—is principally used in hepatic disorders and jaundice. This herbal drug is also attributed with diverse pharmacological properties including antiallergic, antiasthmatic, anti-inflammatory, antimicrobial, antipyretic, antispasmodic, antiviral, bitter tonic, free radical–scavenging, and immunomodulatory effects. It is reported to be useful for treating allergic reactions, bronchial asthma, and ischemic heart disease (Vohora et al. 1972; Vohora 1979; Doshi et al. 1983; Pandey 1989; Balachandran and Govindarajan 2004; Mishra 2004; WHO 2011).

Ethanolic extract (100–400 mg kg^{-1} i.p.) revealed no mortality or behavioral changes up to 400 mg kg^{-1} during the 48-hour period of observation; 1 out of 10 mice in the group treated with 500 mg kg^{-1} i.p. died after 36 hours of injection. Rats treated at these doses showed no discernible symptoms and effect on spontaneous motor activity. Clinical studies (at recommended doses) exhibited no adverse effects in patients who were treated for liver disease and jaundice. Large doses are purgative and should be used with caution in patients with loose motions and in pregnant women. The bitter taste of their herbal drug may induce nausea and vomiting in sensitive individuals. This tendency can be marked by consuming the drug along with honey or sweet syrup. Jaundice with complications (e.g., itching, weight-less, edema) should be properly investigated and treated under medical supervision. In a clinical study (random, placebo controlled, double blind) in 72 patients with bronchial asthma, the results were not very encouraging. No significant reduction in clinical exacerbation, need for bronchodilators, or improvement in pulmonary function was observed. Fifty patients dropped at different stages of the study. Significant side effects were noted in 10 patients (vomiting: 4, anorexia: 3, itching: 2, and giddiness: 1). As this herbal drug is frequently used (singly and in formulations), further investigations are warranted to probe the safety aspects (Doshi et al. 1983; Balachandran and Govindarajan 2004; WHO 2011).

57. *Piper longum* L. (Pipal, Pippali). Fruits and leaves contain lignans, piperamides, dihydropiperonaline, cyclohexane derivatives, and so on. The plant is attributed with adaptogenic, anthelmintic, antiallergic, antiphlegmatic, antitussive, appetizing, decongestant, expectorant, immunomodulatory, tonic, and vasorelaxant properties. It is used for the treatment of various diseases: diarrhea, chronic dysentery, asthma, Alzheimer's disease, ischemic heart disease, male reproductive dysfunction, and so on. (Dhar 1989; Balachandran and Govindrajan 2004; Mishra 2004; WHO 2011).

    Pippali powder is considered to be safe (at the recommended doses) for the treatment of cough and respiratory ailments. While Ayurvedic texts do not mention any toxicity or adverse effects from its use, a WHO monograph (2011) does not advise its long-term use as a single plant; it should be taken with honey and warm water. LD_{50} values have been reported to be 750–800 mg kg^{-1} in mice.

58. *Pongamia pinnata* Merr. (Karanj). Seeds have been reported to possess anthelmintic, antibacterial, antifungal, and insecticidal properties. It is used for the treatment of skin diseases (e.g., eczema, herpes genitalis and herpes zoster, leprosy, leucoderma, ringworm, scabies), gonorrhea, impetigo, pityriasis versicolor, and so on. Petroleum ether extracts demonstrated reversal of cognitive deficit following a two-week treatment in experimental models of Alzheimer's disease. Essential oil is responsible for its medicinal effects (Babu et al. 1995; Mishra 2004; WHO 2011).

59. *Pterocarpus marsupium* Roxb. (Bijsar). Heartwood and gum (Kino) is used for medicinal purpose; its principal actions are hypoglycemic or antihyperglycemic for use in the treatment of diabetes mellitus. Experimental studies in animals and some clinical reports are available to support its utility in this disease. Water kept overnight in tumblers (made from the heartwood of this plant) is claimed to be of value for diabetic patients and in chest pain. While these tumblers are sold in the Indian markets, no scientific studies are available to validate (or refute) the claims of their efficacy. Clinical studies with crude extract and purified principles exhibited encouraging results in NIDDM. Other properties attributed to the plant include hypotensive effects and utility in cardiovascular diseases. Some chemical compounds isolated from the plant are epicatechin, marsupin, massipol, pterostilbine, and pterosupin (Gupta 1963; Mukherjee et al. 1963; Shah 1967; Chaudhury and Vohora 1970b; Vohora 1979; Chakravarthy et al. 1982; Mishra 2004).

While no adverse effects were discernible in clinical studies, antidiabetic activity in some reports was attributed to marked reduction in glucose absorption from the GIT due to the high tannin content in this herbal drug. It is not known whether there was concurrent hindrance to the absorption of other essential nutrients also. This needs to be investigated. Epicatechin (1 g kg^{-1}) was found to be nontoxic in rats (Mukherji et al. 2004; Vohora and Vohora 2005).

60. *Randia dumetorum* Retz. Synonym: *Catunoregm spinosa* Thunb. (Menphol, Emetic Nut). The fruit pulp has been attributed with varied medicinal properties and uses: analgesic, anthelmintic, antifertility, anti-inflammatory, antimicrobial (including antiviral), and emetic effects. It is claimed to be useful for the treatment of poisonous snake bite cases. Some phytochemical and pharmacological studies are available for this herbal drug. Chemical compounds isolated from the plant are 3-β-glucoside, oleonolic acid, saponins, and so on. The latter has been reported to cause irritation of mucus membranes and hemolysis of whole blood and washed blood corpuscles (hemolytic index 1 out of 40,000). It showed emetic and expectorant action in dogs due to its irritant action on the gastric mucosa. Other toxic effects include hypertensive effects in rabbits and dogs (fatal in large doses), lowering of tone and motility of isolated rabbit intestine, and cardiac arrest in perfused heart preparations, and so on (Gupta and Sharma 2007).
61. *Ricinus communis* L. (Eranda, Castor bean). Roots, bark, stems, seeds or beans, and castor oil are used for varied medicinal effects and therapeutic uses. These include its antiarthritic or anti-inflammatory and laxative properties with utility in idiopathic constipation, liver disease, rheumatoid arthritis, and sciatica. Citable reports on controlled clinical trials are not available. These are warranted because the drug is widely used in the above-mentioned conditions in Ayurveda. Phytotoxins present in seeds and ricin (a lectin) in seeds and pods are poisonous. Severe diarrhea and gastrointestinal problems have been reported (Khanna et al. 1969; Chopra et al. 1980; Premila 1989; Mishra 2004; Rates 2001; Saini 2004; Bnouham et al. 2006; Coopman et al. 2009).

   Please also see examples 8, 9, and 25 under Belgium (Coopman et al. 2009), Brazil (Rates 2001), and example 35 under Morocco (Bnouham et al. 2006) in Table 3.1 in Chapter 3.
62. *Semecarpus anacardium* L. (Bhilwa, Marking Nut). Fruits, seeds or kernels, pulp, and bark are used. This herbal drug contains oil, anacardic acid, bhilwanol, and semecarpol. It is attributed with various medicinal properties and uses, for example, anthelmintic, antiarthritic, anti-inflammatory, antimicrobial, antimutagenic, antioxidant, and immunomodulatory properties. The drug retards lipid peroxidation and is used for creating diverse diseases including sciatica pain.

   It is a toxic drug causing irritation during excretion from the urinary tract skin and perianal region. This results in urticaria, rashes, and bleeding. A report claims that it is simple to control these symptoms by administration of antihistaminic drugs. Anacardic acid is considered the toxic principle (Khanna et al. 1969; Vohora 1979; Chopra et al. 1980; Vohora and Wani 1987; Mishra 2004).

63. *Swertia chirata* Roxb. ex Flem. (Chirayata). The whole plant is attributed with diverse pharmacological properties and uses anthelmintic, anti-inflammatory, antimalarial, antipyretic, antitubercular, cholagogue, hepatoprotective, laxative, and stomachic effects with utility in gastrointestinal disorders, nonspecific fevers, liver disease, leishmaniasis, diabetes, and so on (Chopra et al. 1980; WHO 2011). While crude powder generally causes no toxicity, a WHO monograph (2011) advised certain precautionary measures: (1) Patients on other antidiabetic drugs should be given with caution as it may potentiate the hypoglycemic action, (2) medication should be stopped if the fever is not mitigated within a few hours or the symptoms aggravate, (3) patients with very high fever and chronic fever should, preferably, avoid this drug and seek medical advice, and (4) pregnant women should use it with caution.
64. *Symplocos racemosa* Roxb. (Lodhra). Bark and seeds possess varied medicinal properties: anti-inflammatory, antimicrobial, astringent, and styptic effects with utility in treating diverse conditions including gynecological disorders (leucorrhea, menorrhagia, and metrorrhagia). While no adverse effects are observed at recommended doses, some precautions have been recommended: (1) It is advisable to diagnose the cause of leucorrhea before initiating the medication. (2) Overdose or intake on an empty stomach may cause a feeling of heaviness in the abdomen, nausea, and vomiting in persons who are prone to gastrointestinal problems. The symptoms can be controlled by taking a light and liquid diet. (3) Spicy and sour food items (yoghurt) and stress should be avoided. Mental stress aggravates leucorrhea. (4) The dose should be reduced if the menstrual flow is reduced. (5) Long-term use should be avoided in pregnant women. The use by nursing mothers is, however, safe for the baby (WHO 2011).
65. *Syzgium aromaticum* L. (Laung, Lavang, Clove). Dried flower buds and oil are attributed with analgesic, anesthetic, antiseptic, antispasmodic, carminatives, digestive, stomachic, and rubefacient properties and utility in toothache. The drug causes local and systemic adverse effects: (1) Irritant and pungent actions. Care should be taken that the oil, used for treating toothache, does not touch the gum and tongue. Application in deep dental cavities needs caution. If its topical application fails to relieve pain, advice from a dental physician should be taken. (2) Clove is toxic. Oral use in large doses (3.7 g kg^{-1} body weight) may be life-threatening. (3) Undiluted clove oil may cause itching and rashes (even burns), so even local application needs caution. (4) Enough information on safety aspects for its oral use is not available. Such use should, therefore, be avoided especially, in pregnant and nursing women (WHO 2011).
66. *Tachyspermum ammi* (Bishop's Weed). Seeds, leaves, and seed oil are used as a spice and flavoring agents in food, toothpastes, and mouthwashes. It possesses antiseptic, antispasmodic, antioxidant, carminative, digestive, diuretic, and germicidal properties. Excessive use may lead to dehydration,

reduce milk secretion and semen volume, and cause damage to eyes (Bakhru 2001).

67. *Tamarix indica* (Imli, Tamarind). Fruit pulp, leaves, and bark are used in food, culinary preparations, and chutneys. This herbal drug is attributed with antiflatulent, astringent, cooling, digestive, diuretic, and relaxant properties and utility in treating biliary disease. Its sour taste may cause hyperacidity and cough. It is also said to cause sexual weakness (Bakhru 2001).
68. *Taxus baccta* L. (Talispatra, Zarnab). Its leaves are reported to show anticancer, antifertility (anti-implantation), antiseptic, antispasmotic, carminative, cardiotonic, diuretic, emmenagogue, hypotensive, and tranquilizing properties. Plant cell factories constitute an alternative source for anticancer phytochemicals that are biosynthesized in *Taxus* spp. The nature has made this a chosen plant for biotechnological production of *Taxol*, the active principle of this herbal drug (Khanna et al. 1969; Vohora and Kumar 1971; Vohora 1972; Malik et al. 2011).

    The total extract of the leaves showed LD_{50} i.p. value of 5 g kg^{-1} (after 24 hours) and 2.2 kg^{-1} (after 48 hours) in albino mice. Hypotensive action, observed with this extract, was not found to be medicated through cholinergic nerves, vagus, ganglia, carotid receptors, sympathetic, or histaminergic mechanisms. It was attributed to direct depressant action on the CNS and cardiac muscles (Vohora 1972).
69. *Terminalia arjuna* Wight & Arn. (Arjun). The bark is primarily used for the treatment of heart disease in Ayurvedic medicine (angina pectoris, congestive heart disease, coronary artery disease, cardiac arrhythmias, etc.). It is attributed with various pharmacological properties: antiarrhythmic, anticoagulant, antimitogenic, antioxidative, antitumor, antiviral, diuretic, hypolipidemic, thrombolytic, and so on. Considerable experimental and clinical data are available to support its protective and curative value in cardiac ailments. The effects have been probed for its efficacy when used as a single plant and as an ingredient of herbal formulation (Apana and Hartone). The stem bark has shown potent antihypercholesterolemic effects in rabbits. It reduced total cholesterol, low-density lipoprotein, very low density lipoprotein, free fatty acids, phospholipids, and elevated high-density lipoprotein in hypercholesterolemic rabbits. Ethanolic and aqueous extracts exhibited hypotensive, bradycardiac, myocardial depressant, and relaxation of epinephrine-contracted aorta in experimental animals. Stem bark increased endogenous superoxide dismutase and reduced glutathione, catalase, and thiobarbituric acid–reactive substances in the heart and protected the heart from oxidative stress during ischemic–reperfusion injury *in vitro*. Mechanisms of action were also investigated.

    Clinically, the stem–bark extract demonstrated its efficacy in angina and ameliorative effect in hypertension and heart failure cases. The drug is quite safe at doses that are prescribed in traditional systems of medicine. LD_{50} for the crude extract was found to be 2.5 g kg^{-1} i.p. in mice. Rats and rabbits did not show any morality, untoward effects, and histopathological changes in heart, liver, and kidneys in doses up to 10 g kg^{-1} po that are administered for 40 days. Mice appear to be more sensitive to this herbal drug versus rats and rabbits. Clinical studies at optimum dose (1–2 g per day) in patients with coronary artery disease revealed that the drug was well tolerated. While some patients complained of mild gastritis, constipation, and headache, no metabolic, hepatic, or renal toxicity was observed in patients who were given this drug for more than two years (Singh et al. 1982; Pathak et al. 1990; Tiwari et al. 1990; Jain et al. 1992; Dwivedi and Agrawal 1994; Bharoni et al. 1995; Gauthaman and Mishra 2004; Gupta et al. 2005).

70. *Thevetia neriifolia* Juss. ex Staud. Synonym: *T. peruviana* Schem. (Pila Kaner, Yellow Oleander). Kernels or seeds of this ornamental shrub are primarily used as a cardiotonic. All parts of the plant have been used in folk medicine for varied medicinal effects and as an insecticide. Experimental and clinical reports support its beneficial action in cardiac ailments (Arora and Rangaswami 1972; Vohora 1979; Chopra et al. 1980). These have been attributed to glycosidal principles: Peruvoside, Ruvoside, Thevetin, Neriifolin, Cerebrin, and so on.

    All parts of the plant (particularly seeds and latex) may produce toxic effects in man and animals in large doses. These range from diarrhea (bark), abortion, and heart block (seeds and latex). Heart block is produced through significant interference with ATPase-medicated Na, K pump, and electrical conduction mechanisms. A dose of just two seeds is already toxic for adults, and death may follow in ingestion of 8–10 seeds. There are some reports about its suicidal use. Clinical features of its toxicity resemble those of digitalis but with more rapid onset. Ingestion causes burning sensation in mouth, tingling of tongue, dryness of throat, nausea, vomiting, and giddiness. Electrocardiogram changes include atrioventricular block, bradycardia, and depression of the ST segment. Death is preceded by a sharp fall in blood pressure and arrhythmia. Toxic glycosides are considered responsible for its poisonous effects (Arora and Rangaswami 1972; Pahwa and Chatterjee 1990; Rates 2001; Saini 2004).

    Please also see examples 10.3 and 11 under Brazil (Rates 2001) and Canada (Pahwa and Chatterjee 1990) in Table 3.1 in Chapter 3.

71. *Tribulus terrestris* L. (Chotta Gokhru). Its fruits and whole plant are primarily used for diuretic effects. Other medicinal properties that are attributed to this plant include antimicrobial, anti-inflammatory, aphrodisiac, emollient, cytoprotective, and styptic effects. Therapeutic utility has been reported for the treatment of gastrointestinal disorders, renal and cystic stones, and enlarged prostates (Seth and Sethy 1970; Vohora 1979; Chopra et al. 1980; WHO 2011).

    The seeds have exhibited hepatotoxic effects in rats. It is advisable to give deseeded fruits in patients with liver dysfunction who are under medical supervision. Serious urinary symptoms (pus and blood in urine and inability to urinate) should not be treated with this drug. Medical advice should be sought in such cases. A soft digestive diet and abstaining from frequent or strenuous sex are advisable to patients with urinary disorders. Mild joint pain in both arms was found to be the common adverse effect following treatment with a formulation (containing this drug) that is administered to treat benign-type hyperplasia (Talasaz et al. 2010; Sengupta et al. 2011; WHO 2011).

Please also see examples 26 under Iran (Talasaz et al. 2010) in Table 3.1 in Chapter 3.

72. *Tylophora asthmatica* Wight & Arn. Synonym: *Tylophora indica* Herr. (Anantamul). Its leaves contain Tylophorine and Tylophorinine. The plant is attributed to its antiasthmatic and anticancer properties. Considerable experimental and clinical evidence exists for its beneficial use in bronchial asthma and allergic rhinitis (Vohora 1979; Chopra et al. 1980; Mishra 2004; WHO 2011).

    A crossover double-blind study with one leaf daily (chewed and swallowed early in the morning) showed good effect in cases of bronchial asthma and allergic rhinitis following a 4–12-week treatment. Some side effects, for example, sore throat, irritation of mucus membranes of tongue, loss of taste for salt, and/or morning nausea, were observed in 53% of treated patients (Shivpuri et al. 1969).
73. *Vinca rosea* L. Synonym: *Catharanthus roseus* G. Don. (Sada Bahar, Red Periwinkle). The whole plant and flowers are used for medicinal purpose. The principal use of this herbal drug is for the treatment of cancer. Considerable experimental and clinical data are available to support its value in this disease. Its active principles are Vinblastine, Vincristine, and Vindoline. Other chemical constituents isolated from it are reserpine, serpentine, oleonolic acid, α-Amyrin acetate, and so on. In addition to its main action as an anticancer agent, the plant is attributed with antimicrobial (antibacterial, antifungal, antiviral), antidiabetic, antihypertensive, hypolipidemic, insecticidal, and nematocidal effects and utility for treating warts (Chattopadhyay and Das 1990; Gupta and Tandon 2004; Gupta and Sharma 2007). In a clinical study, a paste of fresh leaves (10 g per day in two divided doses) showed encouraging results in diabetic patients (significant decrease in polyurea, polydipsia, weakness, and nonsignificant improvement in other symptoms: polyphagia, cramps on walking, libido, joint pain, and weight loss). No hypoglycemia was observed in healthy volunteers (Goyal and Tiwari 1999). Fifteen patients with malignant pleural effusions were treated with Vincristine (2 mg). Twelve (80%) showed complete resolution of pleural fluid; two exhibited reduced effusions with no need for recurrent aspiration. One patient did not recover and required repeated aspirations (Vidyasagar et al. 1999). Allergic contact dermatitis was reported to be an adverse effect of pollens from this (Kundu et al. 1985; Gupta and Tandon 2004; Gupta and Sharma 2007).
74. *Withania sommifera* Dural. Synonym: *W. ashwagandha* Kaul. (Asgandh, Ashwagandha). The roots are attributed with diverse medicinal properties in Indian systems of medicine. These include adaptogenic, antibacterial, antifungal, antiviral, anticancer, antiaging, antioxidative, aphrodisiac, immunostimulant, tonic (general, cardiac, nervine, and sexual) and tranquilizing

effects. Pharmacological studies have confirmed its immunomodulatory, cytoprotective, and antiaging properties. It is considered a valuable vitalizing drug and is promoted as a male sexual tonic (libido, potency, quality of semen, retentive power, and fertility enhancer) and has been termed as Indian ginseng in a detailed review on the plant. Active compounds isolated from this herbal drug are ashwagandhine, ashwagandhanine, withaferin, and so on. (Vohora 1979; Qureshi et al. 1989; Kulkarni and Dhir 2008b; WHO 2011).

W. somnifera is generally considered safe at recommended doses (up to 9 g per day for four weeks). Larger doses may cause abortion, and so the drug is contraindicated in pregnant women. Intake by nursing mothers is, however, considered safe for the baby. Malaise persisting for more than a week should be properly investigated for the underlying cause. Individuals with hot constitution should be given smaller doses. Concomitant use of alcohol and psychotropic drugs should be avoided (WHO 2011).

Please also see examples 45 under Sri Lanka (Arseculeratne et al. 1985) in Table 3.1 in Chapter 3.

75. *Zaleya gravid* Buch-Ham. (Patthar Chatta). It has been reported to cause diarrhea, abortion, and paralysis (Saini 2004).

4.3 FORMULATIONS

Polypharmacy is the rule, rather than an exception, in Ayurveda and Unani-Tibb (the Greco-Arab system of medicine) practiced on the Indian subcontinent. It is not uncommon to find herbal drug combinations or herbometallic formulations with 20–30 or more ingredients. The exponents of the traditional system of medicine believe that the presence of different ingredients promotes the efficacy (by synergistic action with other ingredients) and reduces the toxicity (by antagonizing the toxic or adverse effects of other components) of the formulation. These claims have not been scientifically validated (or refuted) for most of the formulations. Available data on the subject are limited. Some examples are given as follows.

4.3.1 Heavy Metals

Kumar and Gupta (2012) from the All India Institute of Medical Sciences and Department of AYUSH, Ministry of Health and Family Welfare, Government of India, New Delhi, analyzed a total of 78 Ayurvedic formulations marketed in Delhi for Hg, As, Cd, and Pb content by AA and validated by ICP-MS methods. Their findings are summarized in Table 4.2.

4.3.2 Ayurvedic Formulations

Pant et al. (2012) analyzed Ayurvedic formulations for heavy metal content. They found 17 out of 70 products tested to contain heavy metals (Pb, Hg, and/or As). The products with high heavy metal content are shown in Table 4.3.

TABLE 4.2
Formulations Studied

Herbal	Herbometallic	Metal
56	19	3
	Formulation with Content above WHO Permissible Limits	
Herbal	**Herbometallic**	**Metal**
Pb 11 (19.6%)	10 (52.6%)	3 (100%)
Cd 12 (21.4%)	5 (26.3%)	3 (100%)
Hg, As 3 (5.3%)	1 (5.2%)	3 (100%)

TABLE 4.3
Ayurvedic Formulations with High Metal Content

S. No.	Product	Manufacturer
1	*Bal Champa*	Jairam
2	*Bala Guti*	Zandu
3	*Bala ugathi*	Navjeevan
4	*Balaguti Kesaria*	Kesari Ayurvedic Pharma
5	*Gesari*	Harnarayan Pharmacy
6	*Karela*	Himalaya
7	*Mahadarshan Churna*	Dabur
8	*Mahadarshan Churna*	Zandu
9	*Mahalakshmi Vilas Ras with Gold*	Baidynath
10	*Mahayograj Gugglu with Silver and Makardhwaja*	Baidynath
11	*Navratana Rasa*	Unjha
12	*Safi*	Hamdard
13	*Swarna Mahayograj Gugglu with Gold*	Baidynath

Note: Preparations at S. No. 1–4 are intended for use in infants.

4.3.3 Arogyawardhini Vati (AV)

AV is a traditional Ayurvedic herbometallic formulation used for centuries with claims of efficacy and safety in the treatment of liver disease, jaundice, and skin disorders. It contains the following ingredients: (1) *Picrorhiza kurroa* (Kutki), (2) *Terminalia chebula* (Haritaki), (3) *Terminalia belerica* (Bibhitaka), (4) *Emblica officinalis* (Amlaki), (5) *Shilajit* (Asphalt), (6) *Commiphora mukul* (Guggul), (7) *Ricinus communis* (Eranda), (8) *Azadirachta indica* (Neem), (9) *Shuddha Rasa* (Purified Mercury), (10) *Shuddha Gandhaka* (Purified Sulfur), (11) *Lauh Bhasma* (Calcined Iron preparation), (12) *Abhra Bhasma* (Calcined Mica preparation), (13) *Tamra Bhasma* (Calcined Copper preparation).

Kumar et al. (2012a) carried out a safety evaluation of this herbometallic preparation on the brain, liver, and kidneys of male Wistar rats. AV in doses equivalent up to 10 times of the human dose administered to rats for 28 days did not show discernible toxicological effects on the brain, liver, and kidneys. There were no significant changes in biochemical parameters or histopathology at all doses. This was despite the dose-dependent increase in tissue Hg concentrations. The authors recommended further confirmation for lack of effects after chronic administration as this is more relevant for some neurological effects that are seen with Hg ingestion. AV, which has a history of long usage, proved to be quite safe at recommended doses in this study. It supports its use in humans at such doses and duration.

4.3.4 Arjuna Powder and Arogyavardhini Vati

Kumar et al. (2012b) conducted a double-blind clinical evaluation of combined treatment with *Terminalia arjuna* (Arjuna) powder and *Arogyavardhini Vati* (AV) in 96 patients with dyslipidemia. The study was carried out for seven weeks. The patients received Arjuna powder (5 g twice a day) for the first three weeks followed by AV (500 mg twice a day) for the next four weeks. They found the treatment to be quite effective and safe for dyslipidemia and claimed that the therapy offers significant protection against atherosclerosis and with a reduction in dose and adverse effects of commonly used modern hypolipidemic agents.

4.3.5 Chyawanprash Awaleha (CA)

CA is a polyherbal *Rasayan* (rejuvenating) formulation that is described in *Charak Samhita.* While it contains many ingredients, the primary ingredient is the pulp of *Emblica officinalis* (Amlaki), a rich source of vitamin C. This preparation is popularly used in India and is manufactured by many herbal drug companies including Dabur, Zandu, Baidynath, and Hamdard. It is attributed with immunomodulatory, tonic properties and is used as a health supplement in India and many other countries. Intake during winter is claimed to protect against cough, cold, and respiratory disease. In a controlled clinical study using CA as an adjuvant in the treatment of pulmonary tuberculosis, it was observed that the preparation could augment the recovery process in addition to increased nutritional status of the subjects. The preparation is also reported to exhibit anticarcinogenic effects in mice. A clinical study was undertaken to estimate the radioprotective effect of CA. The study included patients with head and neck cancers who are undergoing radiotherapy. It was observed that CA conferred an excellent protection to the tissues from burning effects of radiation in comparison with controls. The tumor regression rate was also found to be higher in the CA-treated group. The results suggest possible antioxidant effects of CA (Mungatiwar and Phadke 2004). No reports of toxicity or untoward effects of CA are available.

4.3.6 Herbomineral Formulations

In studies carried out by the Indian Institute of Toxicology Research, Lucknow, three herbomineral formulations (*Mahayograj Guggulu*, *Arogyawardhini Vati,* and *Mahalakshmi Vilas Rasa*) were found to be safe (28-day toxicity studies, subchronic studies, and analysis of major tissues for metal retention) (Symposium on Strategies of Safety Study Requirements for Herbal Formulations 2010).

4.3.7 Miscellaneous Formulations and Cosmetics

Researchers from DIPSAR reported (1) the presence of As and Pb in cosmetics; (2) the presence of nicotine in herbal dentifrices; 2 out of 8 brands of tooth powders (*Vicco* and *Musa ka Gul*), and 4 out of 8 tooth powders (*Dabur Red*, *Aredant*, *IPCO*, and *Dentobac*) tested; and (3) the presence of PDE-5 inhibitors in herbal aphrodisiacs (Symposium on Strategies of Safety Study Requirements for Herbal Formulations 2010).

4.3.8 Chaturbhuja Decoction

The Chaturbhuja Decoction is used for indigestion and contains *Ativisha*, *Guduchi*, *Mustoka*, and *Shunthi*. Though the formulation is considered to be traditionally safe at recommended doses, and side effects are most unlikely, a WHO monograph (2011) advised some precautions and safety measures: (1) the presence of *Ativisha* and *Shunthi* may induce symptoms of dryness in the mouth, nervous depression, and tremors following an overdose. (2) Pregnant women should take it under medical supervision. Intake by nursing mothers is, however, safe for the body. (3) If indigestion is induced or aggravated to mental stress, relaxation or rest is advisable.

4.3.9 Trikatu Powder

Trikatu Powder, used for common cold, contains (1) *Zingiber officinale* (Shunthi, Dry ginger rhizome), (2) *Piper nigrum* (Maricha, Black Pepper fruit), and (3) *Piper longum* (Pippali, Long Pepper fruit). It is used in combination with honey to alleviate many diseases, for example, colds, rhinitis, coughs, breathlessness, asthma, dyspepsia, and obesity. The adult dose is 2 g three times a day administered with warm water, milk, or honey. In children, the dose is reduced according to age (125–500 mg thrice a day). *Trikatu* powder may also be taken after boiling it in a cup of milk. The formulation is also attributed with analgesic, anti-inflammatory, carminative, digestive, and expectorant properties.

While no toxic or adverse effects are generally observed, the formulation is "hot" in nature. Larger doses should be avoided as the patient may complain of burning sensation, burning micturition, and so on. In such conditions, it should be stopped immediately, and the patient should be advised to take cooling agents and plenty of milk till the symptoms subside completely. Medication with drug should be avoided in pregnant women. Intake by nursing mothers is, however, safe for the baby (WHO 2011).

4.3.10 Triphala Decoction

The Triphala Decoction used for the treatment of discharge from the eyes and other diseases, contains three myrobalans: fruits of *Amlaki*, *Bibhitaka*, and *Haritaki*. It is a rich source of vitamin C: other chemical principles present in the formulation include carotene, nicotinic acid, riboflavin, glycosides, anthraquinones, polyphenolic compounds, gallic acid, and tannins. It is attributed with anti-inflammatory, decongestant, soothing, and wound-healing properties. No serious adverse effects have been reported in classical literature. A WHO monograph (2011) recommends the same precautions and safety measures: (1) The decoction should always be made fresh, in hygienic conditions using clean utensils, and avoiding dipping of fingers. (2) For washing eyes, the decoction should not be too hot or too cold as such conditions may not give the desired result. (3) Frequency depends on severity of symptoms (normal: twice a day). (4) Warm compresses should be applied to soften and remove crusts from eyelids. (5) If the symptoms do not improve or worsen (decreased vision, itching, thick purulent or greenish discharge) following a one-to-two-week treatment, the drug should be stopped and medical advice sought.

4.3.11 Hepatoprotective Formulations

There are 33 patent herbal formulations available on the Indian market for use in liver ailments. The most popular and widely used preparations among these are Liv-52 (Himalaya Drug Co., Mumbai) and Tefroli (TTK Pharma Pvt. Ltd., Chennai). These are frequently prescribed by allopathic physicians. Though as many as 81 medicinal plants have been used in such formulations, 28 plants are contained in 5 or more formulations. The 10 plants that top the list are shown in Table 4.4.

Pharmacological and toxicity data on most of these plants are available. The formulation Liv-52 seems to be most extensively investigated from all aspects: pharmacological, toxicological, and clinical, with very encouraging results. It has reportedly exhibited protective and regenerative effects on hepatic parenchyma, anabolic, diuretic, choleretic, and appetite-improving properties in experimental animals and has had good results in cases of anorexia, childhood cirrhosis, infective hepatitis, chronic active hepatitis, adult cirrhosis, jaundice, and so on. The preparation (available in the form of tablets and syrup) has made a place for itself in the treatment of liver diseases, which do not respond favorably to modern medicine (Vohora and Amin 1993). Liv-52 has been evaluated for safety in pregnancy and location (Charan 2012). No reports of adverse effects are available.

REFERENCES

Abraham, K., Wohrlin, F., Lindtner, O., Heinmeyer, G., and Lampen, A. 2010. Toxicology and risk assessment of coumarin: Focus on human data. *Mol Nutr Food Res* 54: 228–238.

Adedapo, A.A., Omoloya, O.A., and Ohore, O.G. 2007. Studies on the toxicity of an aqueous extract of the leaves of *Abrus precatorius* in rats. *Onderstepoort J Vet Res* 74: 31–36.

Agarwal, R.C. and Bhide, S.V. 1987. Biological studies on toxicity of chillies in Balb/c mice. *Indian J Exp Biol* 36: 391–396.

Agarwal, R.C. and Bhide, S.V. 1988. Histological studies on chilli (Capsaicin) in Syrian hamsters. *Indian J Exp Biol* 26: 377–382.

Agrawal, S.S. 2010. Safety assessment and monitoring of marketed medicine including cosmetics and herbal drugs. Available at http://www.isrpt.co.in/Downloads/presentations_2010/day1pre14conf2010.ppt, accessed on 12/4/2012.

Ahmad, M., Yousuf, S., Khan, M.B., Abdullah, M.N.H., Ahmad, S., Ansari, M.A., Ishrat, T., Agrawal, A.K., and Islam, F. 2006. Attenuation by *Nardostachys jatamansi* induced Parkinsonism in rats: Behavioral, neurochemical and immunochemical studies. *Pharmacol Biochem Behav* 83: 150–160.

Akbarsha, M.A., Subramoniam, A., and Madhavachandran, V.M. 2004. Male reproductive dysfunction. In: *Scientific Basis for Ayurvedic Therapies* (ed. Mishra, L.C.) pp. 459–477, CRC Press, Boca Raton, Florida.

Ali, M., Bhagat, A., and Gupta, J. 1995. Comparison of anti-inflammatory activity Curcumin-analogues. *Indian Drugs* 32: 502–505.

Ali, S., Ansari, K.A., Jafry, M.A., Kabeer, H., and Diwakar, G. 2000. *Nardoctachys jatamansi* protects against liver damage induced by thioacetamide in rats. *J Ethnopharmacology* 71: 359–363.

Alvarez-Perea, A., Garcia, A.P., Hernandez, A.L.R., De Barrio, M., and Baeza, M.L. 2010. Utricaria due to Aloe vera: A new sensitizer? *Ann Allergy Asthma Immunol* 105: 404–405.

Amresh, G., Singh, P.N., and Rao, C.V. 2008. Toxicology screening of traditional medicine Laghupatha (*Cissampelos pareira*) in experimental animals. *J Ethnopharmacol* 16: 454–460.

Ang-Lee, M.K., Moss, J., and Yuan, C.S. 2001. Herbal medicines and perioperative care. *JAMA* 286: 208–216.

Anto, R.J., George, J., Babu, K.V.D., Rajasekharan, K.N., and Kuttan, R. 1996a. Anti-mutagenic and anti-carcinogenic activity of natural and synthetic curcuminoids. *Mutat Res* 370: 127–131.

Anto, R.J., Kuttan, G., Babu, K.V.D., Rajasekharan, K.N., and Kuttan, R. 1996b. Anti-inflammatory activity of natural and synthetic curcuminoids. *Amla Res Bull* 16: 73–77.

Anto, R.J., Kuttan, G., Babu, K.V.D., Rajasekharan, K.N., and Kuttan, R. 1998. Anti-inflammatory activity of natural and synthetic curcuminoids. *Pharm Pharmacol Commun* 4: 103–106.

Arora, R.B. 1965. *Nardostachys Jatamansi: A Chemical Pharmacological and Clinical Appraisal.* Special Report Series No. 51, Indian Council of Medical Research, New Delhi.

Arora, R.B. and Rangaswami, S. 1972. *Peruvoside and Other Cardiotonic Glycosides of Thevetia neriifolia Juss: Chemical, Pharmacological and Clinical Studies.* CCRIMH Monograph Series No.1, Thomson Press India Ltd., New Delhi.

Arseculeratne, S.N., Gunatilaka, A.A., and Panshokke, R.G. 1985. Studies on medicinal plants of Sri Lanka Part 14. Toxicity of some traditional medicinal herbs. *J Ethnopharmacology* 13: 323–335.

Ashafa, A.O.T., Orekoya, L.O., and Yakubu, M.F. 2012. Toxicity profile of ethanolic extract of *Azadirachta indica* stem bark in Wistar rats. *Asian Pac J Trop Med* 2: 811–817.

Babu, T.D., Kuttan, G., and Padikkala, J. 1995. Cytotoxic and anti-tumour properties of certain taxa of Umbelliferae with special reference to *Centella asiatica* (L.) *Urban J Ethnopharmacol* 48: 53–57.

Bagiga, M.S., Bhat, H.P., Baliga, B.R.V., Wilson, R., and Paltty, P.L. 2011. Phytochemistry, traditional uses and pharmacology of *Eugenia jambolana* Lam (black plum): A review. *Food Res Internat* 44: 1776–1789.

Bajaj, A.K., Govil, D.C., and Bhargova, S.N. 1982. Dermatitis due to plants. *Indian J Dermatol Venerol Leprol* 48: 268–270.

Bakhru, H.K. 2001. *Indian Spices and Condiments as Natural Healers.* Jaico Publishing House, Mumbai.

Balachandran, P. and Govindarajan, R. 2004. Hepatic disorders. In: *Scientific Basis for Ayurvedic Therapies* (ed. Mishra, L.C.), pp. 231–253.

Balaji, S. and Champaran, B. 2010. Toxicity prediction from compounds from turmeric (*Curcuma longa* L.). *Food Chem Toxicol* 48: 2951–2959.

Balasubramanyam, K., Varier, R.A., Altaf, M., Swaminathan, V., Sidappa, N.B., Ranga, U., and Kunda, T.K. 2004. Curcumin, a novel p300/CREB-binding protein specific inhibitor of acetyltransferase, suppresses the acetylation of histone/non-histone protein and histone acetyltransferase-dependent chromatin transcription. *J Biol Chem* 279: 51163–51171.

Bandhopadhyay, I., Ganguly, C.K., Chakraborty, T., Bhattacharya, K., and Banerjee, R.K. 2002. Gastroprotective effect of Neem (*Azadirachta indica*) bark extract: Possible involvement of H^+ K^+, ATP-ase inhibition and scavenging of hydroxyl radical. *Life Science* **71**: 2845–2865.

Banerjee, A. and Nigam, S.S. 1978a. Anti-microbial efficacy of the essential oil of *Curcuma longa*. *Indian J Med Res* 68: 864–866.

Banerjee, A. and Nigam, S.S. 1978b. *In vitro* anthelmintic activity of essential oil derived from various species of *Curcuma* Linn. *Sci Cult* 414: 503–504.

Bansal, R., Ahmad, N., and Kidwai, J.R. 1981. Effect of oral administration of *Eugenia jambolana* seeds and chlorpropamide on blood glucose level and pancreatic Cathepsin B in rats. *Indian J Biochem Biophys* 18: 377–380.

Bhagat, G.R., Kaji, B.C., and Chudgar, N.P. 1993. Study of fifty cases of air pollen in contact with respiratory allergy. *Indian J Intern Med* 3: 206–209.

Bhagat, M. and Purohit, A. 2001. Anti-fertility activity of various extract of *Curcuma longa* in male albino rats. *Indian Drugs* 38: 79–81.

Bharoni, A., Ganguly, A., and Bhargava, K.D. 1995. Salutary effect of *Terminalia arjuna* in patients with severe refractory heart failure. *Indian J Cardiol* 49: 191–196.

Bhatia, A., Singh, G.B., and Khanna, N.M. 1964. Effect of *Curcumin* oil on histamine-induced gastric ulceration. *Indian J Exp Biol* 2: 158–160.

Bhatia, R.Y.P. 1974. Specificity of some plant lectins in differentiation of animal blood. *Forensic Sci* 4: 47.

Bhawani Shankar, T.N., Shantha, N.V., Ramesh, H.P., Murthy, I.A.S., and Murthy, V.S. 1980. Toxicity studies on turmeric (*Curcuma longa*): Acute toxicity in rats, guinea pigs and monkeys. *Indian J Exp Biol* 18: 73–75.

Bhide, S.V., Agarwal, R.C., and Sundaram, V. 1993. Biological studies on carcinogenicity of chilli extract in mice, rats and hamsters in different conditions. *Amla Res Bull* 13: 61–72.

Bidwai, P.P., Wangoo, D., and Bhullar, N. 1990a. Anti-spermatogenic action of *Celastrus paniculatus* seed extract in the rats with reversible changes in the liver. *J Ethnopharmacology* 28: 293–303.

Bidwai, P.P., Wangoo, D., and Sharma, V. 1990b. Effect of polar and semipolar compounds from the seeds of *Celastrus pamiculatus* on liver and kidneys of rats. *Fitoterapia* 61: 417–424.

Bnouham, M., Mehrfour, F.Z., Elachoui, M., Legssyer, A., Mekhfi, H., Lamnouer, D., and Ziyyat, A. 2006. Toxic effects of some plants used in Moroccan traditional medicine. *Moroccan J Biol* 2–3: 21–30.

Boeke, S.J., Boersma, M.G., Alink, G.M., Van Loon, J.J.A., Van Huis, A., Dicke, M., and Rietiene, I.M.C.M. 2004. Safety evaluation of neem (*Azadirachta indica*) -derived pesticides. *J Ethnopharmacol* 44: 25–41.

Bone, K. 1996. *Gymnema sylvestre*. In: *Clinical Applications of Ayurvedic and Chinese Herbs: Monograph for Western Practitioners* (ed. Bone, K.), Phytotherapy Press, Warwick, Queensland.

Brahamchari, H.T. and Augustu, K.T. 1961. Hypoglycemic agents from Indian indigenous plants. *J Pharm Pharmacol* 13: 381–385.

Chakravarthy, B.K., Gupta, S., and Gode, K.D. 1982. Functional β-cell regeneration in the islets of pancreas in alloxan-induced diabetic rats by (–) epicatechin. *Life Sci* 31: 2693–2698.

Chandra Shekhar, J., Sandhya, S., Vinod, K.R., Banji, D., Sudhakar, K., and Chaitanya, RSNAKK. 2012. Plant toxins: Useful and harmful effects. *Hygeia: Journal for Drugs & Medicines* 4: 79–90.

Chandrasekaran, C.V., Deepak, H.B., Thiyagarajan, S., Sangli, G.K., Deepak, M., and Agarwal, A. 2011. Dual inhibitory effect of *Glycyrrhiza glabra* on COX and LOX products. *Phytomedicine* 18: 278–284.

Charan, T.S. 2012. Manager, Customer Service, the Himalaya Drug Co. Bangalore, personal communication, December 14.

Chattopadhyay, S.P. and Das, P.K. 1990. Evaluation of *Vinca rosea* in the treatment of warts. *Indian J Dermatol Venerol Leprol* 50: 107–108.

Chaubul, P.D. and Gadve, S.B. 1984. Study of pollen allergy in Kolhapur during monsoon. *Indian J Chest Dis* 26: 439–444.

Chaudhary, N., Deshmukh, V.K., and Aiman, R. 1961. Effect of indigenous plants on glucose absorption in mice. Studies with detannated extracts of the bark of *Ficus bengalensis Indian J Physiol Pharmacol* 5: 18–21.

Chaudhury, R.R. and Vohora, S.B. 1970a. Indigenous antifertility plants. In: *Advances in Research on Indian Medicine* (eds. Udupa, K.N., Chaturvedi, G.N., and Tripathi, S.N.), pp. 197–222, Banaras Hindu University, Varanasi.

Chaudhury, R.R. and Vohora, S.B. 1970b. Plants with possible hypoglycemic activity. In: *Advances in Research on Indian Medicine* (eds. Udupa, K.N., Chaturvedi, G.N., and Tripathi, S.N.) pp. 52–75, Banaras Hindu University, Vanarasi.

Chemexcil Board of Editors. 1992. *Selected Medicinal Plants of India: A Monograph on Identity, Safety and Clinical Usage*. Compiled by Bharatiya Vidhya Bhavan's S.P.A.R.C. Mumbai for Chemexcil, Basic Chemical, Pharmaceuticals and Cosmetics Export Promotion Council, Bombay.

Chinnoy, N.J., Multani, A.A., and Chakravorty, N. 1997. Effect of papaya on micro-environment and sperm metabolism in epididymus of rats. *J Med Aromat Plants* 19: 422–426.

Chitra, S., Sabitha, K., Nalini, N., and Menon, V.P. 1994. Influence of red chilli on lipids and bile acids in different tissues in experimental colon cancer. *Indian J Exp Biol* 32: 793–796.

Chopra, R.N., Nayar, S.L., and Chopra, I.C. 1980. *Glossary of Indian Medicinal Plants*. Council of Scientific and Industrial Research, New Delhi.

Choudhary, D., Chandra, D., and Kale, R.K. 1999. Modulation of radio response of glyoxalase system by Curcumin. *J Ethnopharmacol* 64: 1–7.

Coopman, V., De Leeuw, M., Cordonnier, J., and Jacobs, W. 2009. Suicidal death after injection of castor bean extract. (*Ricinus communis* L.). *Forensic Sci Int* 189: e 13–e 20.

Cravotto, G., Boffa, L., Genzine, L., and Garella, D. 2010. Phyto-therapeutics: An evaluation of the potential of 1000 plants. *J Clin Pharm Ther* 35: 11–48.

Dadhaniya, P., Patel, C., Muchhara, J., Bhadja, N., Mathuria, N., Vachhani, K., and Soni, M.G. 2011. Safety assessment of a solid lipid curcumin preparation: Acute and sub-chronic studies. *Food Chem Toxicol* 49: 1834–1842.

Dalsaniya, V.D., Patel, D.V., Patel, H.V., and Bhagat, G.R. 1999. Allergen skin tests. Analysis of results in patients of bronchial asthma. *Indian Pract* 52: 402–406.

Dandiya, P.C. 1999. Traditional herbal drugs: Efficacy, safety and toxicity. In: *Toxicology and Environmental Health* (eds. Vohora, S.B. and Agrawal, V.P.), Ed 1: 137–145, Society of Biosciences, Jamia Hamdard and Asiatech Publishers Inc., New Delhi.

David, P.K. and Peter, C.T. 1970. Toxicity of *Abrus precatorius* in domestic fowl. *Kerala J Vet Sci* 1: 125–127.

Deshpande, S.S., Ingle, A.D., and Maru, G.B. 1998. Chemopreventive efficacy of Curcumin-free aqueous extract in 7, 12-dimethylbenz(a)anthracene-induced mammary tumerogenesis. *Cancer Lett* 123: 35–40.

Dhar, K.L. 1989. Recent advance in the chemistry of *Piper*. In: *Research and Development of Indigenous Drugs* (eds. Dandiya, P.C. and Vohora, S.B.), pp. 81–87, Institute of History of Medicine and Medical Research, New Delhi.

Dixit, V.P., Khanna, P., and Bhargava, S.K. 1978. Effect of *Momordica charantia* fruit extract on the testicular function of dog. *Planta Medica* 34: 280–284.

Doshi, V.B., Shetye, M., and Mahashur, A.A. 1983. *Picrorhiza kurroa* in bronchial asthma. *J Postgrad Med* 29: 89–93.

Drugs and Cosmetics Act 1940 and Rules 1945, Ministry of Health and Family Welfare, Govt. of India.

Dwivedi, S. and Agrawal, M.P. 1994. Anti-anginal and cardioprotective effect of *Terminalia arjuna,* an indigenous drug, in coronary artery disease. *J Asso Physicians India* 12: 287–290.

EU Food Law. 2006. Cinnamon flavoured foods recalled over coumarin levels, Nov.

Faghihi, G. and Radan, M. 2011. Side effects of herbal drugs in dermatology. *J Cosmetic Dermatol Sci Appl* 1: 1–3.

Garg, S.K. 1974. Effect of *Curcuma longa* (rhizomes) on fertility in experimental animals *Planta Med* 26: 225–227.

Gauthaman, K. and Mishra, L.C. 2004. Ischemic heart disease. In: *Scientific Basis for Ayurvedic Therapies* (ed. Mishra, L.C.), pp. 511–534, CRC Press, Boca Raton, Florida.

George, P. 2011. Concerns regarding the safety and toxicity of medicinal plants: An overview. *J Appl Pharmaceut Sci* 1: 40–44.

Godkar, P.B., Narayanan, P., and Bhide, S.V. 1996. Hypocholesterolemic effect of turmeric extract on Swiss mice. *Indian J Pharmacol* 28: 171–174.

Goyal, D.K. and Tiwari, S.K. 1999. Study of Sadabahar (*Vinca rosea* Linn.) in Madhumeha (diabetes mellitus). *J Res Ayur Siddha* 20: 93–100.

Grieco, A., Miele, L., Pompili, M., Vecchio, F.M., Grottagliando, I., and Garburrini, G. 2009. Acute hepatitis caused by a lipid-lowering product: When alternative medicine is no alternative at all. *J Hepatol* 50: 1273–1277.

Gupta, A.K. and Gupta, S.N. 1994. Allergens in airway diseases. *Ann Natl Acad Sci (India)* 30: 113–114.

Gupta, A.K. and Sharma, M (eds.). 2007. *Reviews on Indian Medicinal Plants*, Vol. 5, Indian Council of Medical Research, New Delhi.

Gupta, A.K. and Tandon, N (eds.). 2004. *Review on Indian Medicinal Plants*. Indian Council of Medical Research, New Delhi, 1: 18–38.

Gupta, A.K., Tandon, N., and Sharma, M (eds.). 2005. *Quality Standards of Indian Medicinal Plants*, Vol. 2–3. Indian Council Medical Research, New Delhi.

Gupta, A.K., Tandon, N., and Sharma, M (eds.). 2006a. *Quality Standards of Indian Medicinal Plants*. Indian Council of Medical Research, New Delhi, 1: 45.

Gupta, A.K., Tandon, N., and Sharma, M. 2006b. *Quality Standards of Indian Medicinal Plants*. Indian Council of Medical Research, New Delhi, 4: 25.

Gupta, A.K., Tandon, N., and Sharma, M (eds.). 2008. *Quality Standards of Indian Medicinal Plants*. Indian Council Medical Research, New Delhi, 6: 122,123.

Gupta, L., Tayal, G., Agarwal, S.S., and Bhardwaj, S.L. 1974. Toxicity studies on *Cannabis indica*. *J Res Indian Med* 9: 102–104.

Gupta, S.K. 2002. *Emergency Toxicology: Management of Common Poisons*. Narosa Publishing House, New Delhi.

Gupta, S.S. 1963. Effect of *Gymnema sylvestre* and *Pterocarpus marsupium* on glucose tolerance in albino rats. *Indian J Med Sci* 17: 501–505.

Gupta, S.S. 1989. Indigenous drugs in bronchial asthma. In: *Research and Development of Indigenous Drugs* (eds. Dandiya, P.C. and Vohora, S.B.), pp. 168–173, Institute of History of Medicine and Medical Research, New Delhi.

Hameed, A. and Vohora, S.B. 2001. *Indian System of Medicine Skin Disease: A Herbo-Mineral Approach*. CBS Publishers and Distributers, New Delhi.

Hasan, S.S. and Kushwaha, A.K.S. 1987. Chronic effect of Datura seeds on brain of albino rats. *Jap J Pharmacol* 44: 1–6.

Homma, M., Oka, K., Nitisuma, J., and Itoh, H. 1993. Pharmacokinetic evaluation of traditional Chinese herbal remedies. *Lancet* 341: 1595.

Hussain, A. 1989. Importance of medicinal plants in health care programme of the world. In: *Research and Development on Indigenous Drugs* (eds. Dandiya, P.C. and Vohora, S.B.), pp. 21–33, Institute of History of Medicine and Medical Research, New Delhi.

ICMR. 2008. Quality standards of Indian medicinal plants. *Indian Council of Medical Research*, New Delhi, 5: 61.

Iyer, G.V. and Narendranath, M. 1975. A preliminary report on myocardial manifestations of *Cerebra odolam* poisoning. *Indian J Med Res* 63: 312–314.

Jagetia, G.C. and Rajnikant, G.K. 2004. Role of Curcumin, a naturally occurring phenolic compound of turmeric in accelerating the repair of excision wound in mice whole body exposed to radiation. *J Surg Res* 120: 127–138.

Jain, A.K., Tezuka, H., Kada, T., and Tomata, I. 1987. Evaluation of genotoxic effect of turmeric in mice. *Curr Sci* 56: 1005–1006.

Jain, J.P., Naqvi, S.M.A., and Sharma, K.D. 1990. A clinical trial of volatile oil of *Curcumalonga* Linn. (Haridra) in cases of bronchial asthma (Tamak Swasa). *J Res Ayur Siddha* 11: 20–30.

Jain, V., Poonia, A., Agrawal, R.P., Kochar, D.K., and Mishra, S.N. 1992. Effect of *Terminalia arjuna* in patients of angina pectoris (a clinical trial). *Indian Med Gaz (New Series)* 36: 56–59.

Jamir, N.S. 1989. Some interesting medico-botany used by AO: A Naga tribe. In: *Research and Development of Indigenous Drugs* (eds. Dandiya, P.C. and Vohora, S.B.), pp. 259–264, Institute of History of Medicine and Medical Research, New Delhi.

Jha, V.K., Sundaramma, M., Mishra, S.K., and Joshi, M. 1975. Clinical studies with some airborne pollens and dust around Varanasi. *Indian J Chest Dis* 17: 107–112.

Joseph, L.J., Deshpande, U.R., and Samuel, A.M. 1999. The role of thyroid hormones in protective effect of turmeric extract in d-galactosamine HCl-induced hepatic injury in rats. *Thyroidol Clin Exp* 11: 1–6.

Joshi, J., Ghasias, S., Vaidya, A., Vaidya, R., Kamat, D.V., Bhagwat, A.N., and Bhide, S. 2003. Early human safety study of turmeric oil (*Curcuma longa* oil) administered in healthy volunteers. *J Asso Phys India* 51: 1053–1060.

Kakkad, P. 2011. Toxicological assessment of herbal materials: Current status and progress in India. In: *Proceedings of One-Day Symposium on Safety and Risk Assessment Approaches for Materials of Herbal Origin*, Bangalore, September 3, 2010. Toxicology International Supplement, August.

Kapadia, G.J., Patel, B.D., Chug, E.B., Ghosh, B., and Pradhan, S.N. 1976. Carcinogenicity of *Camelia sinensis* (tea) and some tannin-containing folklore medicinal herbs administered subcutaneously in rats. *J Natl Cancer Inst* 57: 207–209.

Kar, A., Chaudhuray, B.K., and Bandhopadhyay, N.G. 2003. Comparative evaluation of hypoglycemic activity of some Indian medicinal plants in alloxan-diabetic rats. *J Ethnopharmacol* 84: 105–108.

Katiyar, C.K. 2012. Safety evaluation of Ayurvedic drugs, regulatory requirements, issues and perspectives. In: *Herbal Products: Regulatory Aspects* (ed.), T.N. Medical College, Mumbai. Available at http://www.tnmcnair.com/docs/herbal%20products%20-%20seminar/Day%201/Session%20II/CK%20Katiyar_Safety%20Evaluation%20of%20Ayurvedic%20Drugs_Regulatory%20Requirements,%20Issues%20and%20Perspectives.pdf.

Kaur, P., Kaur, S., Kumar, N., Singh, B., and Kumar, S. 2009. Evaluation of antigenotoxic activity of isoliquitinaposide from *Glycyrrhiza glabra* L. *Toxicology* 23: 680–686.

Kaur, P., Kumar, M., Singh, B., Kumar, S., and Kaur, S. 2012. Amelioration of oxidative stress induced by mutagens and COX-2 inhibitory activity Umbelliferone isolated from *Glycyrrhiza glabra* L. *Asian Pacific J Trop Biomed* 2(1 Supp): 5120–5126.

Kavli, G. and Volden, G. 1984. Phytodermatitis. *Photodermatology* 1: 65–75.

Khanna, U. and Chaudhury, R.R. 1968. Antifertility screening of plants. Part I. Investigations on *Butea monosperma* (Lam) Kuntz. *Indian J Med Res* 56: 1575–1580.

Khanna, U., Garg, S., Vohora, S.B., Walia, H.B., and Choudhury, R.R. 1968. Anti-fertility screening of plants. Part II: Effect of six indigenous plants on early pregnancy in albino rats. *Indian J Med Res* 57: 237–244.

Khanna, U., Garg, S.K., Vohora, S.B., Walia, H.B., and Chaudhury, R.R. 1969. Antifertility screening of plants. Part II. Effect of six indigenous plants on early pregnancy in rats. *Indian J Med Res* 57: 237–244.

Kino, P.M. and Pai, K.V. 1965. Cardiotoxic effects of *Cerebra odollam. Indian Heart J* 17: 263–270.

Koley, J., Koley, B.N., and Maitra, S.R. 1973. Effect of drinking tea, coffee and caffeine on work performance. *Indian J Physiol Allied Sci* 31: 96–106.

Koley, J., Dey, C.D., Koley, B.N., and Mitra, S.R. 1977. Alpha rhythm and physical fatigue: Effect of tea, coffee and caffeine. *Indian J Physiol Allied Sci* 34: 36–44.

Kulkarni, S.K. and Dhir, A. 2008a. On the mechanism of anti-depressant-like action of berberine chloride. *Europ J Pharmacol* 589: 163–172.

Kulkarni, S.K. and Dhir, A. 2008b. Review: *Withania somnifera*: An Indian ginseng. *Progress Neuropsychoparmacol Behav Psychiatr* 32: 1093–1105.

Kumar, G. and Gupta, Y.K. 2012. Monitoring of mercury, arsenic, cadmium and lead in Ayurvedic formulations marketed in Delhi by flame AAS and confirmation by ICP-MS. *Food Addit Contam Part B Surveill* 5: 140–144.

Kumar, G., Srivastava, A., Sharma, S.K., and Gupta, Y.K. 2012a. Safety evaluation of Ayurvedic medicine, *Arorgyavardhini Vati*, on brain, liver and kidney in rats. *J Ethnopharmacology* 140: 151–160.

Kumar, G., Srivastava, A., Sharma, S.K., and Gupta, Y.K. 2012b. Safety efficacy evaluation of Ayurvedic treatment (*Arjuna* powder and *Arorgyavardhini Vati*) in dyslipidemia patients: A pilot prospective cohort clinical study. *AYU* 33: 1–5.

Kundu, S., Chanda, S., and Kundu, I. 1985. Clinical investigations on pollen-grains from Calcutta. *Trans Bose Res Inst* 48: 21–28.

Kuntali, K., Pillai, M.V., and Krishna Dass, K.V. 1970. Poisoning with *Cerebra odolam. Indian Heart J* 22: 373–377.

Lau, B.H.S., Tadi, P.P., and Tosk, J.M. 1990. *Allium sativum* (Garlic) and cancer prevention. *Nutr Res* 10: 937–948.

Lodha, R. and Kabra, S.K. 2004. Bronchial asthma. In: *Scientific Basis for Ayurvedic Therapies* (ed. Mishra, L.C.), pp. 209–229, CRC Press, Boca Raton, Florida.

Luyckx, V.A. 2012. Nephrotoxicity of alternative medicine practice. *Adv Chronic Kidney D* 19: 129–141.

Lyle, N., Gomes, A., Tapas, S., Munshi, S., Paul, S., Chatterjee, S., and Bhattacharya, D. 2009. The role of anti-oxidant properties of *Nardostachys jatamansi* in alleviation of symptoms of chronic fatigue syndrome. *Behav Brain Res* 202: 285–290.

Maiti, R.N., Manickam, M., Ray, A.B., and Goel, R.K. 1997. Effect of Withafastuosin E on gastric mucosal offensive and defensive factors in rats. *Indian J Exp Biol* 35: 751–753.

Malik, M.K., Kumar, A., and Bhalla, B.P.R. 1983. Cannabis ototoxicity. *Indian J Otolaryngol* 35: 74.

Malik, S., Cusido, R.M., Mirjalili, M.H., Moyano, E., Palteon, J., and Bonfill, M. 2011. Production of anti-cancer drug taxol in *Taxus baccata* suspension cultures. A review. *Process Biochem* 46: 23–34.

Manickam, M.W., Padma, P., Chansouria, J.P.N., and Ray, A.B. 1997. Evaluation of anti-stress activity of Withafastuosin D, a withanolide of *Daturafastuosa. Phytother Res* 11: 384–385.

Mishra, D.S., Sharma, R.P., and Soni, B.K. 1996. Toxic hemaglutinating properties of *Abrus precatorius* L. *Indian J Exp Bio* 4: 161–163.

Mishra, L.C. 2004. (ed.). *Scientific Basis for Ayurvedic Therapies*. CRC Press, Boca Raton, Florida, USA.

Mishra, L.C. and Adra, T. 2004. Diabetes mellitus (*Madhumeha*) In: *Scientific Basis for Ayurvedic Therapies* (ed. Mishra, L.C.), pp. 101–121, CRC Press, Boca Raton, Florida.

Mishra, L.C. and Singh, R.H. 2004. Parkinson's disease (*Kampa Vata*). In: *Scientific Basis for Ayurvedic Therapies* (ed. Mishra, L.C.), pp. 453–458, CRC Press, Boca Raton, Florida.

Mishra, R., Billiore, K.V., Yadav, B.B.L., and Chaturvedi, D.D. 1992. Some ethno-medicinal plants—Lore from Ajmer forest division (Rajasthan). *J Econ Tax Bot* 16: 421–424.

Mitra, S.K. and Rangesh, P.R. 2004a. Diarrhoea (*Atisara*) and dysentery (*Pravahika*). In: *Scientific Basis for Ayurvedic Therapies* (ed. Mishra, L.C.), pp. 371–392, CRC Press, Boca Raton, Florida.

Mitra, S.K. and Rangesh, P.R. 2004b. Irritable colon. (*Grahni*). In: *Scientific Basis for Ayurvedic Therapies* (ed. Mishra, L.C.), pp. 355–369, CRC Press, Boca Raton, Florida.

Mitra, S.R., Mukherjee, S.P., and Koley, B.N. 1967. Effect of tea drinking on muscular efficiency with special reference to relief on physiological fatigue. *J Exp Med Sci* 11: 56–62.

Mukherjee, S.N., De, U.N., and Mukherjee, B. 1963. Contribution in the field of diabetes research in the last decade. *Indian Med Gez* 3: 97–100.

Mukherjee, S.K., Saxena, A.M., and Shukla, G. 2004. *Progress of Diabetes Research in India during 20th Century*. National Institute of Science Communication, New Delhi.

Mungantiwar, A.A. and Phadke, A.S. 2004. Immunomodulation: Therapeutic strategy through Ayurveda. In: *Scientific Basis for Ayurvedic Therapies* (ed. Mishra, L.C.), pp. 63–81, CRC Press, Boca Raton, Florida.

Nagabhusan, M. and Bhide, S.V. 1985. Mutagenicity of chilli extract and capsaicin in short-term tests. *Environ Mutagen* 7: 881–888.

Nalini, N., Sabitha, K., Viswanathan, P., and Menon, V.P. 1998. Influence of spices on bacterial (enzyme) activity in colon cancer. *J Ethnopharmacol* 62: 15–24.

Naqvi, S.A.H., Khan, M.S.Y., and Vohora, S.B. 1991. Anti-bacterial, antifungal and anthelmintic investigations on Indian medicinal plants. *Fitoterapia* 62: 221–228.

Ossenkoppele, P.M., van der Sluis, W.G., and van Vloten, W.A. 1991. Phototoxicity following the use of *Ammimajus* fruit for vitiligo. *Ned J jdscr Geneeskd* 135: 478–480.

Pahwa, R. and Chatterjee, V.C. 1990. The toxicity of yellow oleander (*Thevetia neriifolia* Juss) seed kernals in rats. *Vet Hum Toxicol* 32: 561–564.

Panda, S., Bharti, S., and Kar, A. 2002. *Emblica officinalis* and *Bauhinia purpurea* in regulation of thyroid function and lipid peroxidation in male mice. *J Herbs Spices Med* 10: 1–8.

Pandey, V.N. 1989. A clinical and experimental study on Ayurvedic drug Kutki (*Picrorhiza kurroa*): A major constituent of *Arogyavardhini*. In: *Research and Development of Indigenous Drugs* (eds. Dandiya, P.C. and Vohora, S.B.), pp. 122–130, Institute of History of Medicine and Medical Research, New Delhi.

Panigrahi, G.B. and Rao, A.B. 1983. Influence of caffeine on arecoline-induced SCE in mouse bone marrow cells *in vivo*. *Mutation Res* 122: 347–353.

Pant, K.K., Katiyar, C.K., and Gupta, A. 2012. Safety aspects of Ayurvedic drugs. In: *Evidence-Based Practice in Complementary Alternative Medicine* (ed.), pp. 211–232, Sanjeev Rastogi, Springer Berlin Heidelberg.

Parida, M.M., Upadhyay, C., Pandya, G., and Jana, A.M. 2002. Inhibitory potential of neem (*Azadirachta indica* Juss) leaves on Dengu virus2 replication. *J Ethnopharmacol* 79: 273–278.

Parmar, V. 2005. Herbal medicine: Its toxic effects and drug interactions. The Indian Anaesthetists Forum. Available at http://www.theiaforum.org, ISSN 0973-0311.

Pasricha, J.S. and Kanwar, A.J. 1978. Substances causing contact dermatitis. *Indian J Dermatol Venereol Leprol* 44: 264–268.

Patel, V.K. and Bhatt, V.H. 1988. Folklore therapeutic indigenous plants in periodontal disorders in India (review, experimental and clinical approach). *Int J Clin Pharmacol Ther Toxicol* 26: 176–184.

Pathak, S.R., Upadhyaya, L., Singh, R.H., Dubey, G.P., and Udupa, K.N. 1990. Effect of *Terminalia arjuna* W. & A. On autocoidal and lipid profiles of rabbits. *Indian Drugs* 27: 221–227.

Paul, R., Prasad, M., and Shab, N.K. 2011. Anti-cancer biology of *Azadirachta indica* (neem): A mini review. *Cancer Biol Ther* 12: 467–476.

Porchezhian, E. and Dobriyal, R.M. 2003. An overview of advances in *Gymnema sylvestre* leaf extract: Chemistry, pharmacology and patents. *Pharmazie* 58: 5–9.

Prabhu, K.S., Saldanha, K., and Pattabiraman, T.N. 1984. Natural enzyme inhibiters: A comparative study of the action of legume inhibitors in human and borine pancreatic proteinases. *J Sci Food Agric* 35: 314–321.

Pramod, G.C. and Udupa, K.K. 1970. Role of *Cissus quadrangularis* on fracture healing. In: *Advances in Research in Indian Medicine* (eds. Udupa, K.N., Chaturvedi, G.N., and Tripathi, S.N.), pp. 163–196, Banaras Hindu University, Varanasi.

Premila, M.S. 1989. Safety of herbal drugs, the need for caution. In: *Research and Development of Indigenous Drugs* (eds. Dandiya, P.C. and Vohora, S.B.), Institute of History of Medicine and Medical Research, New Delhi.

Qureshi, S., Shah, A.H., Tariq, M., and Angel, A.M. 1989. Studies on herbal aphrodisiacs used in Arab system of medicine. *Amer J Clin Med* 17: 57–62.

Raman, A. and Lau, C. 1996. Anti-diabetic properties and phytochemistry of *Momordica charantia* L. *Phytomedicine* 2: 349–362.

Rao, V.S., Rao, A., and Karanth, K.S. 2005. Anti-convulsant and neurotoxic profile of *Nardostachys jatamansi* in rats. *J Ethnoparmacology* 102: 351–356.

Rasheed, A.S., Venkatraman, S., Jayverra, K.N., Fazil, A.M., Yashodha, K.J., Aleem, M.A., Mohammed, M., Khaja, Z., Ushari, B., Pradeep, H.A., and Ibrahim, M. 2010. Evaluation of toxicological and anti-oxidant potential of *Nardostachys jatamansi* in reversing haloperidol-induced catalepsy in rats. *Int J General Med* 3: 127–136.

Rates, S.M.K. 2001. Review: Plants as source of drugs. *Toxicon* 39: 603–613.

Rathi, S.S., Grover, J.K., Vikrant, V., and Biswas, N.R. 2002. Prevention of experimental diabetic cataract by Indian Ayurvedic plant extracts. *Phytother Res* 16: 774–780.

Renter, H.D. 1995. *Allium sativum* and *Allium ursinum*: Part 2. Pharmacology and medical application. *Phytomedicine* 2: 73–91.

Saha, J.C. and Kalyansundaram, S. 1962. Studies on pollen allergy in Pondicherry. Part I. Survey of potentially allergenic plants. *Indian J Med Res* 50: 881–888.

Saini, D.C. 2004. Ethno-phytotoxicologial studies in Siddhi district of Madhya Pradesh. In *Ethnomedical Plants* (eds. Trivedi, P.C. and Sharma, N.K.), Chapter 14, pp. 192–225, Pointer Publishers, Jaipur.

Salim, S., Ahmad, M., Zafar, K.S., Ahmad, A.S., and Islam, F. 2003. Protective effect of *Nardostachys jatamansi* in rat cerebral ischemia. *Pharmacol Biochem Behav* 74: 481–486.

Saper, R.B., Kales, S.N., Paquin, J., Burns, M.J., Eisenberg, D.M., Davis, R.B., and Phillips, R.S. 2004. Heavy metal content of Ayurvedic herbal medicine products. *JAMA* 292: 2868–2873.

Saroja, S., Jayasree, A., and Annapurni, S. 1995. Screening tests for the presence of mutagens, co-mutagens. *Indian J Nutr Dietet* 32: 165–168.

Sateesh, R., Prakash Kumar, R., Joe, J.C., Nair, P.K.K., and Rao, G.R. 1994. *Cassia* pollen grains of India. *Int J Allergy Immunol* 103: 280–285.

Sathe, M., Bhatia, H.M., and Purandare, N.M. 1970. Studies on seed extracts II. Interaction of erythrocytes of various animal species with agglutinine (lectins). *Indian Vet J* 47: 648–656.

Saxena, R.C., Pradhan, R., and Singh, N. 1989. Some herbal drugs of clinical utility. In: *Research and Development of Indigenous Drugs* (eds. Dandiya, P.C. and Vohora, S.B.), pp. 314–321, Institute of History of Medicine and Medical Research, New Delhi.

Sen, S., Agarwal, K., Mukherjee, A., and Sharma, A. 1994. Potentiation by coffee of the frequency of chromosomal aberrations induced by chronic fenfluramine in mice. *Drug Chem Toxicol* 17: 113–124.

Sengupta, G., Hazra, A., Kundu, A., and Ghosh, A. 2011. Comparison of *Murraya koenigii* and *Tribulus terrestris*-based oral formulation versus Tamsulosin in the treatment of benign prostatic hyperplasia in aged (>50 years): A double-blind randomized controlled trial. *Clinical Therapeutics* 33: 1943–1952.

Seth, U.K. and Sethy, V.H. 1970. Indigenous diuretics. In: *Advance in Research on Indian Medicine* (eds. Udupa, K.N., Chaturvedi, G.N., and Tripathi, S.N.), pp. 1–55, Banaras Hindu University, Varanasi.

Shah, D.S. 1967. A preliminary study of the hypoglycemic action of heartwood of *Pterocarpus marsupium* Roxb. *Indian J Med Res* 55: 66–70.

Shah, S., Gabriella, N., Ghosh, D., Fieskova, D., Capek, P., and Ray, B. 2011. Structural features an *in vivo* anti-tussive activity of water extract polymer from *Glycyrrhiza glabra*. *Int J Biol Macromol* 48: 634–638.

Shah, S.A. and Vohora, S.B. 1990. Boron enhances anti-arthritic effects of Garlic oil. *Filoterapia* 61: 121–126.

Shankar, R., Dogra, R.K.S., Gupta, B.N., and Clerk, S.H. 1978. Effect of Hemp dust (*Cannabis sativa* Linn.) on lungs and lymph nodes of guinea pigs. *Indian J Exp Biol* 16: 671–674.

Sharma, S.K. and Singh, A.P. 2012. *In vitro* antioxidant and free radical scavenging activity of *Nardostachys jatamansi* DC. *J Acupunt Meridian Stud* 5: 112–118.

Sharma, M.H., Singh, W.M., Singh, K.R., and Singh, N.I. 2000. Flowering calendar of known allergenic plant growing in Thanbol district, Manipur. *Indian J Aerobiol* 13: 5–9.

Sharma, M., Tandon, N., and Gupta, A.K. (eds.) 2009. *Reviews on Indian Medicinal Plants*, Vol. 8, pp. 302–390, Indian Council of Medical Research, New Delhi.

Shivpuri, D.N. and Dua, K.L. 1963a. Allergy to papaya tree (*Carica papaya* Linn.). *Ann Allergy* 21: 139–144.

Shivpuri, D.N. and Dua, K.L. 1963b. Studies on pollen allergy. Part IV. Clinical investigations. *Indian J Med Res* 51: 68–74.

Shivpuri, D.N., Menon, M.P., and Prakash, D. 1969. A crossover double-blind study on *Tylophora indica* in the treatment of asthma and allergic rhinitis. *J Allergy* 43: 145–150.

Shobi, V. and Goel, H.C. 2001. Protection against radiation-induced conditioned taste aversion by *Centella asiatica*. *Physiol & Behaviour* 73: 19–23.

Singh, B. and Rastogi, R.P. 1972. Chemical examination of *Picrorhiza kurroa* Benth. Part VI: Reinvestigation of Kutkin. *Ind J Chem* 10: 29–32.

Singh, R.G., Usha, and Udupa, K.N. 1982. Clinical and etiological profile of Bhang (*Cannabis sativa* leaves) poisoning in and around Varanasi. *Clinician* 46: 84–91.

Singh, T.U., Kumar, D., Tandon, S.K., and Mishra, S.K. 2009. Inhibitory effects of essential oils of *Allium sativum* and *Piper longum* on spontaneous muscular activity of liver fluke *Fasciola gigantica*. *Exptl Pathol* 123: 302–308.

Singh Gautam, D.N., Singh, P.N., and Mehrotra, S. 1999. Comparative study of processed (*Shodhit*) and unprocessed seeds of Gunja. *Abrus precatorius* L. *Nat Prod Sci* 5: 127–133.

Sirsat, S.M. and Khanolkar, V.R. 1960. Subcutaneous fibrosis of palate in diet-preconditioned Wistar rats. *Arch Pathol* 70: 171–179.

Sivaraman, R., Shah, Z.A., and Vohora, S.B. 2003. Investigations on Dimagheen: A polyherbal formula used in Unani medicine for effects on learning and memory. *Indian Drugs* 40: 160–165.

Sivaraman, R., Shah, Z.A., Gilani, R.A., and Vohora, S.B. 2008. Nootropic activity of Syrup Shankhpushpi: An Ayurvedic polyherbal formulation. In: *Recent Progress in Medicinal Plants: Phytopharmacology and Therapeutic Values* (eds. Singh, V.K., Govil, J.N., and Sharma, R.K.), vol. 19, pp. 81–89, Studium Press LLC, Houston, Texas 77072, USA.

Sridhara, S., Singh, B.D., Kumar, L., Verma, J., Gaur, S.N., and Gangal, S.V. 1995. Antigenic and allergenic relationship among airborne grass pollens in India. *Ann Allergy Asthma Immunol* 75: 72–79.

Srimal, R.C. and Dhawan, B.N. 1973. Pharmacology of diferuloyl methane (Curcumin), a non steroidal anti-inflammatory agent. *J Pharm Pharmacol* 25: 447–450.

Sumathi, S. and Pattabiraman, T.N. 1976. Natural plant enzyme inhibitors. Part II. Protease inhibition of seeds. *Indian J Biochem Biophys* 13: 52–56.

Sundaravalli, N., Raju, R.B., and Krishnamurthy, K.A. 1982. Neem oil poisoning. *Indian J Pediats* 49: 357–359.

Symposium on Strategies for Safety Study Requirements for Herbal Formulations and 30th Annual Conference of Society of Toxicology. 2010. December 9–11, Jamia Hamdard, New Delhi.

Talasaz, A.H., Abbasi, M.R., Abkhiz, S., and Dasti-Khavidaki, S. 2010. *Tribulus terrestris*-induced severe nephrotoxicity in a young healthy male. *Nephrol Dial Transplant* 25: 3792–3793.

Tiwari, A.K., Gode, J.D., and Dubey, G.P. 1990. Effect of *Terminalia arjuna* on lipid profiles of rabbits fed hypercholesterolaemic diet. *Int J Crude Drug Res* 28: 43–47.

Tongia, A., Tongia, S.K., and Dave, M. 2004. Phytochemical determination and extraction of *Momordica charantia* fruits and its hypoglycemic potentiation of oral hypoglycemic drugs in diabetic mellitus (NIDDM). *Indian J Physiol Pharmacol* 48: 241–244.

Tripathi, S.N. 1970. Comparative study of rheumatic arthritis: A disease entity, etio-pathogenesis and treatment. In: *Advances in Research in Indian Medicine* (eds. Udupa, K.N., Chaturvedi, G.N., and Tripathi, S.N.), pp. 223–267, Banaras Hindu University, Varanasi.

Verma, G., Dua, V.K., Agrawal, D.D., and Atul, P.K. 2011. Anti-malarial activity of *Holarrhena antidysentrica* and *Viola canescens* plants traditionally used against malaria in Garhwal Region of north-west Himalaya. *Malaria J* 10: 10–14.

Vidyasagar, M.S., Ramanujan, A.S., Fernades, D.J., Koteshwararao, K., Jadhav, G.K., Hospet, C.S., Setharamaiah, T., Vidyasar, S., and Subrahamanium, K. 1999. Vincristine (*Vinca alkaloid*) as sclerosing agent in malignant plural effusions. *Acta Oncol* 38: 1017–1020.

Vijaylakshmi, G. and Singh, S. 1975. Effects of Cannabis resin on human mitotic chromosome. *J Anat Soc India* 24: 71–73.

Vishwanathan, N. and Joshi, B.S. 1983. Toxic constituents of Indian plants. *Curr Sci* 52: 1–8.

Vohora, D.V. and Mishra, L.C. 2004. Alzheimer's disease. In: *Scientific Basis for Ayurvedic Therapies* (eds. Mishra, L.C.), pp. 411–426, CRC Press, Boca Raton, Florida.

Vohora, S.B. 1972. Studies on *Taxus baccata* II. Pharmacological investigations of the total extract of leaves. *Planta Medica* 22(1): 59–65.

Vohora, S.B. 1974. The Keshari Dal controversy: Useful food or dreadful poison? *Science Reporter,* CSIR, New Delhi.

Vohora, S.B. 1979. *Research on Medicinal Plant in India: Effort and Achievements (1947–1976).* Thesis Ph.D (Science Policy), Jawaharalal Nehru University, New Delhi.

Vohora, S.B. 1989. Research on medicinal plants in India: A review of reviews. *Curare* 12: 16–22.

Vohora, S.B. and Amin, M.M.W. 1993. Some commonly used Ayurvedic and Unani Medical preparations. In: *The Pharmacists Year Book* (eds. Dandiya, P.C., Khar, R.K., and Gurbani, N.K.), pp. 80–83, CBS Publisher and Distributors, Delhi.

Vohora, S.B. and Khan, M.S.Y. 1977. Pharmacological studies on *Lepidium sativum* Linn. *Indian J Physiol Pharmacol* 21: 118–120.

Vohora, S.B. and Kumar, I. 1971. Studies on *Taxus baccata*. I: Preliminary phytochemical and behavioral investigations. *Planta Medica* 20: 100–107.

Vohora, S.B. and Vohora, D. 2005. *Anti-Diabetic Plants: Current Status Report*. Prepared for the INCLEN Trust, Hamdard University, New Delhi.

Vohora, S.B. and Wani, H. 1987. A review of Indian anti-inflammatory plants. *Herba Hungarica* 26 (1): 73–86.

Vohora, S.B., Kumar, I., Naqvi, S.A.H., and Afaq, S.H. 1972. Pharmacological investigation on *Picrorhiza kurroa* (KUTKI) with special reference to its choleretic and antimicrobial properties. *Indian J Pharm* 34: 17–19.

Vohora, S.B., Rizwan, M., and Khan, J.A. 1973. Medicinal uses of common Indian vegetables. *Planta Medica* 23: 381–393.

Wahi, S.P. and Chunekar, K.C. 1964. Pharmacological studies on *Gymnema sylvestre*. *J Sci Res Banaras Hindu Univ* 15: 205–209.

WHO. 2011. *Traditional Herbal Remedies for Primary Health Care*. World Health Organization, Regional Office for South-East Asia, New Delhi.

Witzel, D.A., Dollahite, J.W., and Jones, L.P. 1978. Photo-sensitization on sheep fed *Ammi majus* (Bishop's Weed) seed. *Am J Vet Res* 39: 319–320.

Zafarullah, M. and Vohora, S.B. 1980. *Proceedings of the Seminar on Bars (Leucoderma)*. pp. 125–131, Central Council for Research in Unani Medicine, New Delhi.

5 Regulatory Aspects*

5.1 INTRODUCTION

Chaudhury (1992) classified countries into three categories to discuss the regulatory aspects: (1) countries with heritage in the use of traditional systems of medicine long before the development of modern medicine and where herbal medicine continues to be used as a form of medicine at present, for example, India and China; (2) countries where the modern system of medicine is well developed, but traditional folk medicine, without teaching or textbooks, has always been practiced to some extent, for example, Germany and the United Kingdom; and (3) countries where herbal remedies were not used to any appreciable extent in the past, but in recent times due to their use by migrant populations and a growing interest in herbal remedies by local residents, for example, Canada and Australia. He briefly described the situations that are prevailing in Australia, Brazil, Canada, China, Germany, India, Japan, the United Kingdom, and the United States.

The legal situation regarding herbal preparations varies from country to country. In some countries, phytomedicines are well established, whereas in others, they are regarded as food supplements, and therapeutic claims are not allowed. Developing countries, though, largely depend on traditional herbal medicines (particularly in rural areas), and have poor or vague regulatory laws. The World Health Organization (WHO) has published guidelines on safety monitoring of herbal medicines for a pharmacovigilance system. Despite the growing interest in the safety of herbal medicines, national surveillance systems to monitor and evaluate adverse reactions associated with herbal medicines are rare, even among more than 70 member states that are participating in the WHO International Drug Monitoring Program. Moreover, there is lack of effective communication on this subject at all levels. A recent WHO survey showed that around 90 countries, less than half of WHO's member states, currently regulate herbal medicines, and even a smaller proportion have systems in place for the regulation or qualification of providers of herbal medicines. The current situation is grim with disparities in regulations between countries. This has serious implications for the distribution of such products (WHO 2004).

The WHO International Regulatory Cooperation for Herbal Medicines (IRCH), a global network of regulatory authorities responsible for herbal medicines, was established in 2006 with the mission to promote health and safety through improved regulation for such remedies. Six meetings of IRCH countries were held in Beijing (China), Kuala Lumpur (Malaysia), Montreal (Canada), Dubai (United Arab

* At the time of the writing of this book the *Report of the WHO Survey* (WHO 2005) was in the process of being updated. Readers will be able to refer to it once it is completed: *National Policy and Regulation of Traditional Medicine: Report of the Second WHO Global Survey*, Geneva, WHO (in preparation).

Emirates), Guangzhou (China), and Curitiba (Brazil) from 2006 to 2012. While it is difficult to estimate precisely how far IRCH has succeeded in its mission, it is heartening to note that its membership has increased; from 13 in 2006 to its current membership of 30 as of January 2014. These include 27 countries (Argentina, Armenia, Australia, Brazil, Brunei Darussalam, Canada, Chile, China, Cuba, Ghana, Hungary, India, Indonesia, Japan, Malaysia, Mexico, Oman, Pakistan, Peru, Portugal, the Republic of Korea, Saudi Arabia, Tanzania, the United Arab Emirates, the United Kingdom, and the United States) and three regional bodies: the Association of Southeast Asian Nations (ASEAN), the European Medicines Agency (EMA), and the Latin American Parliament (WHO 2006–2014).

Most individual herbal medicinal products are licensed nationally by member states. The process of licensing and information on such preparations was harmonized across the European Union (EU) through its directives in 2001 and 2004, the Herbal Medicinal Products Committee (HMPC), and *European Pharmacopoeia* (EP). The monographs in the EP provide requirements for herbal substances and preparations. The HMPC addresses quality, safety, and clinical efficacy issues. It also identifies priority for coverage as a monograph or a list entry that is related to single herbal drugs or combinations (Ekor 2013).

Using input from the above mentioned and other sources, a brief summary of the status in 73 countries is given in the following sections (Chaudhury 1992; Homma et al. 1993; Ang-Lee et al. 2001; GCP Guidelines 2001; Rates 2001; Chaudhury and Rafei 2002; WHO 2004; Drugs and Cosmetics Act 1940 and Rules 1945 [Amendments 2005, 2010]; Bandarnayake 2006; Symposium 2010; George 2011; Katoch and Kumar 2012; Human Medicine Regulation 2012; Ekor 2013; Rivera et al. 2013; UNESCO 2013; Verma 2013).

5.2 ANGOLA

While the National Office on Traditional Medicine (TM)/Complementary and Alternative Medicine (CAM) was established under the direction of the Ministry of Health, and an expert committee was constituted in 1998, the regulatory picture is dismal. According to a WHO survey, there was no regulation of herbal medicines in Angola. Herbal medicines were classified as over-the-counter (OTC) medicines but sold in special outlets by licensed practitioners and in markets. Though a national pharmacopoeia and a national monograph on herbal medicines were in development, there was no information available on manufacturing requirements or safety requirements for herbal drugs. No registration system for these medicines was available, and there was no information on the postmarketing surveillance (PMS) of these drugs (WHO 2005).

5.3 ANTIGUA AND BARBUDA

These countries were involved in a project on economic botany of underexploited tropical plants in October 1983. The activities included the development of herbaria, databases, medicinal use, phytochemical analysis, and so forth. There were no existing regulations on traditional medicines (WHO 1998).

5.4 ARGENTINA

For the marketing authorization of a new medicinal product in Argentina, it is generally required to have a laboratory and fulfill the legal requirements of Article 27 of Law No. 16463. There is no difference between herbal medicines and chemical drugs. When the active principle is described in the pharmacopeia, national approval is given by the *National de Farmacologia y Bromatologia*, which is responsible for regulating drugs and food. If the active principle is not described in the pharmacopoeia, a premonograph may be submitted with an application for approval by the institute. The rules or criteria vary in different provinces. In 1993, a regulation for registration and commercialization of medicinal plants was published by the health ministry of one of the provinces. There is a national pharmacopoeia and 899 monographs (WHO 1998).

5.5 ARMENIA

The regulation of herbal medicines in Armenia began in 1998 through a national drug law that also regulates pharmaceuticals. Herbal medicines are regulated as a separate category and as dietary supplements. There is no national pharmacopoeia, and the *British Herbal Pharmacopoeia* is considered, there is a national monograph (*Armenian National Formulary for Herbal Medicines* 2001), but WHO monographs are also being referred to. There are 130 registered herbal medicines and 60 herbal dietary supplements, which are sold in pharmacies as OTC medicines without restriction. The safety regulatory requirements for herbal medicines are similar to conventional pharmaceuticals. The National Center for Postmarketing Surveillance of Medicines was established in 1997 and included reporting for herbal medicinal products (WHO 2005).

5.6 AUSTRALIA

In the State of Victoria (where herbal remedies have been in use for more than 50 years), all herbal drugs sought to be sold in the state have to be registered. All applications are considered by an expert advisory committee, which evaluates them for quality, safety, and efficacy. The toxicity of herbal preparations governs its scheduling and labeling under the Drug Poison and Controlled Substances Law rather than being a prime consideration as a therapeutic substance. Therapeutic goods for human use, which are imported or manufactured in Australia (including traditional medicines), need to be registered under the Therapeutic Goods Act (TGA) of 1989. Alternative medicines are allowed to enter into the Australian Register of Therapeutic Goods following recommendations by the Traditional Medicines Evaluation Committee, which is appointed by the minister. It consists of six to nine members who are experts in clinical practice or teaching of alternative medicine, pharmacy with experience in pharmacognosy, or plant toxicology; the manufacturers of alternative medicines; or medical practitioners or have the qualifications and experience in clinical pharmacology. TGA 1993 appears to be more liberal. While there are special regulations on the expression of quantity or proportion of active ingredients, herbal medicine does not require evidence of efficacy provided that it is not poison on schedule (WHO 1998; Zeman 2014).

Musgrave (1998) surveyed 121 herbal remedies that were advertised and sold on Internet sites from Australia. It was found that (1) nearly one in five such remedies was not licensed under the Australian Register for Therapeutic Goods (ARTG), (2) overall, 60% of the remedies listed ingredients that permitted their license and did not match those that were on the labels, and (3) there was a significant variation in concentration of the content or substitution or addition of active ingredients or a combination of all these discrepancies. The TGA does not regulate herbal medicines from overseas Internet sites. The researcher purchased a subset of these medicines from health food stores and pharmacies. It was found that 53% of these did not comply with their official ARTG listing. The author cited a previous small survey by TGA where 90% of medicines were found to be noncompliant with one or more regulations, and only 71% of surveyed medicines had manufacturing or quality issues; within the ARTG, a remedy may be listed or registered. Registered medicines can make strong therapeutic claims and need convincing evidence backup (biological and toxicological studies and clinical trials). Listed medicines can only make general claims, but it should be ensured that such remedies contain only low-risk ingredients. A random audit by TGA in 2009–2010 again revealed no improvement (90% noncompliance with regulations). Despite these adverse reports, it is surprising to note that the Complementary Healthcare Council maintains that Australia is recognized as one of the highest regarded and most tightly regulated systems in the world. The system relies critically on the cooperation of the herbal medicine sponsors. Failure to remove noncompliant medicines from the market indicates a lack of such cooperation.

5.7 AUSTRIA

In the European Union, a centralized procedure where companies submit a single marketing authorization to the EMA, and a mutual recognition procedure where approval by a single regulatory body is accepted by other countries exists for the registration of herbal products (Abbott 2014). The provisions for the simplified registration of the Traditional Herbal Medicinal Products Directive (THMPD) in the EU can be found in Chapter 2a of Directive 2001/83/EC as amended by Directive 2004/24/EC in April 2004 (Peschel 2014). The simplified registration is acceptable only where the herbal medicinal product relies on long medicinal use in the EU (at least 30 years). Medicinal use outside the EU should be taken into account only if the medicinal product has been used within the EU for at least 15 years (van Galen 2014). The mutual recognition procedure plays an essential role within the European harmonization process for medicinal products, avoiding the need for repeated procedures in the member states (Austria, Belgium, Czech Republic, Denmark, Estonia, Finland, France, Germany, Greece, Hungary, Ireland, Italy, the Netherlands, Poland, Portugal, Romania, Slovakia, Spain, Sweden, the United Kingdom, and others) (Peschel 2014).

Herbal medicine regulations were introduced in Austria in 1989. According to a WHO survey, herbal medicines in Austria were regulated as prescription and OTC medicines and were sold in pharmacies and in special outlets. The *Austrian Pharmacopoeia* and *European Pharmacopoeia* are used and are both legally binding. Regulatory requirements for manufacturing include adherence to information in pharmacopoeias and monographs, and the same rules for good manufacturing practice (GMP) and safety requirements applies as for conventional pharmaceuticals. There was, however, no registration system for herbal drugs, and no national postmarketing system existed for these medicines at that time (WHO 2005).

5.8 AZERBAIJAN

The Republic of Azerbaijan has the same laws and regulations that apply to conventional pharmaceuticals and herbal medicines. Herbal medicines were regulated as OTC medicines, dietary supplements, health foods, and as a separate category. There was no national pharmacopoeia, and herbal monographs were under development. Manufacturing requirements were the same as for conventional pharmaceuticals. No control mechanism exists to ensure compliance with these requirements. There was a registration system for herbal medicines; these were also included in the national essential drugs list, and a PMS system was planned (WHO 2005).

5.9 BANGLADESH

A national policy on TM/CAM was formulated in 1995, and the national program was launched in 1998. The national office under the Ministry of Health had existed since 1990. The same laws and regulations applied to herbal medicines as for conventional pharmaceuticals. Herbal medicines were regulated as prescription and OTC medicines and could be sold with medical, nutrient content, and structure or function claims. The Bangladesh national formularies on Unani and Ayurvedic medicines are legally binding. Regulatory requirements include adherence to pharmacopoeias and monographs and general rules of GMP and safety as for other pharmaceuticals. There was a registration system for herbal medicines, but the information on the number of registered products was not available. There was no PMS system for these drugs (WHO 2005).

5.10 BELARUS

The Republic of Belarus was developing a national policy on TM/CAM. The laws and regulations were the same for herbal drugs and conventional pharmaceuticals. Herbal medicines were regulated as OTC medicines and as dietary supplements. Manufacturing and safety requirements included adherence to pharmacopoeias and monographs and GMP rules as for other drugs. Special requirements of demonstrating traditional use without harmful effects and reference to documented scientific research on similar products are required. The registration procedures require reports on safety. According to a WHO survey, there were 300 registered herbal medicines at that time. PMS exists for herbal medicines (WHO 2005).

5.11 BELGIUM

In Belgium, national policy, and laws and regulations on TM/CAM were introduced in 1999. Regulatory requirements were essentially the same as for conventional drugs. They are sold as prescription and OTC medicines and dietary supplements. Claims may be made about them under law. The *European Pharmacopoeia* is the national pharmacopoeia, and adherence to the same is the regulatory requirement. There is no registration system. However, the PMS system that has been in place since 1990 includes adverse drug reaction (ADR) reporting of herbal medicines (WHO 2005).

5.12 BENIN

The regulations governing TM/CAM in the Republic of Benin were adopted in 2001. Herbal medicines are classified only as OTC medicines and can be sold with claims. There is no national pharmacopoeia or monograph, no regulatory requirements for manufacturing or safety assessment, and no PMS system for herbal drugs (WHO 2005).

5.13 BHUTAN

Bhutan has a traditional system of medicine known as *Sowa Rigpa*. The national policy on TM/CAM is currently being developed, though TM was formally recognized as an integral part of the national healthcare delivery system way back in 1967. The Institute of Traditional Medicine Services was established under the Ministry of Health in the same year. The regulations regarding the manufacture of herbal medicine are akin to the GMP rules for other pharmaceuticals. In 1998, there were 103 herbal medicines listed in the essential category. There were no safety assessment requirements, control mechanisms, and restrictions on the sale of herbal products. No national pharmacopoeia existed in the Kingdom of Bhutan. The general perception is that TM/CAM is good (WHO 2005).

5.14 BOTSWANA

The Republic of Botswana does not have a national policy, laws or regulations, or programs, or national institutes on TM/CAM. While an expert committee was established in 2001, no national pharmacopoeia, monograph, manufacturing, or safety regulations exist (WHO 2005).

5.15 BRAZIL

The WHO (traditional medicine division) recognizes that centuries-old use of certain plants should be taken into account as proof of their efficacy. However, the total acceptance of plant-derived drugs can only occur if the same criteria of efficacy, safety, and quality control are applied to herbal drugs as required for conventional drugs. Details regarding pharmacologically active plant compounds are an essential requirement for standardization of herbal formulations (Rates 2001). In Brazil, the rules for registration of herbal medicines existed since 1967. The two public policies have been available since 2006: the National Policy on Integrative and Complementary Practices in the Public Health System and the National Policy on Medicinal Plants and Herbal Medicines. Herbal medicines are currently regulated by a Brazilian health surveillance agency (Agência Nacional de Vigilância Sanitária [ANVISA]), created in 1999, which is responsible for all health products and the rules and guidelines of which have been recently revised. According to the revised guidelines, herbal products have been categorized as herbal medicines (HMs) and traditional herbal products (THPs). While HMs are required to prove their safety and efficacy through clinical trials, THPs need to demonstrate long-term safety and efficacy in humans (Carvalho et al. 2014). The requirement for registration includes a demonstration of safety, efficacy, and quality control, the methodologies of which have been described in detail in pharmacopeias that are recognized by ANVISA. There were 382 HMs that were registered in Brazil (Carvalho et al. 2014). *Farmacopéia Brasileira* is Brazil's pharmacopoeia, which includes their national monographs, and is legally binding (Fan et al. 2012).

5.16 BULGARIA

The Drug Act 1995 (amended in 2001) closely resembles EU regulations of the pharmaceutical sector. The law does not distinguish between medicinal products that are made from chemical substances and those from plants or natural substances. Proof of quality, safety, and efficacy is a precondition for regulation of all medicines. Registration for borderline products was simplified in 1996 for nonprescription drugs with restricted claims and that are suitable for self-medication (WHO 1998). No national pharmacopoeia or monograph existed. The *European Pharmacopoeia* and European Scientific Cooperative on Phytotherapy (ESCOP) monographs were used, but they are not legally binding. Regulatory or safety requirements for manufacturing were the same as for conventional pharmaceuticals. These were controlled through manufacturing authorization inspections and traditional use without demonstrated harmful effects. Herbal medicines are sold as prescription or OTC medicines. A total of 113 herbal medicines were registered according to a WHO survey, and there is a PMS system (WHO 2005).

5.17 BURKINA FASO

Herbal medicines are sold as OTC or dietary supplements. There is no national pharmacopoeia or monograph, and Burkina Faso uses the *Senegalese Pharmacopoeia* (1974) and *African Pharmacopoeia* (1985) instead, but they are not legally binding. While a traditional medicine pharmacopoeia service (administered by the Ministry of Health, Ouagadougou) was founded, and an expert committee was established in 2000, the regulation scenario was dismal. The regulatory requirements for safety were limited to special requirements that included traditional use without harmful effects and documented research on similar products. No control mechanism existed for these regulations. There was no registration system. Though a PMS system was present, it was without proper control (WHO 2005).

5.18 BURUNDI

While a national office (under the Ministry of Health) and an expert committee on TM/CAM were established in 2002, no national policy on laws and regulations, institutes, national pharmacopoeia or monographs, registration system, and controls over manufacturing or safety seem to exist. Herbal medicines are sold OTC for self-medication. Other pharmacopoeias are used, but they are not legally binding. Herbal medicines are sold in special outlets by registered practitioners without restrictions. Generally, the same GMP rules are followed for conventional drugs and herbal remedies (WHO 2005).

5.19 CAMEROON

A national office (administered by the Ministry of Health) was established in 1995, and the School of Traditional Medicine started in 2001 in the Republic of Cameroon. National regulations of herbal medicines were introduced in 1998 and were partly the same as for other drugs. Herbal drugs in Cameroon are sold as prescription or OTC medicines, and claims may be made about them. The *European Pharmacopoeia* is used. A registration system existed, and 10 medicines were registered. A postmarketing system was under development. One herbal medicine was included on the national essential drugs list that was established in 1999 (WHO 2005).

5.20 CANADA

All natural health products (NHPs) in Canada are regulated by Health Canada under the Food and Drugs Act and the Natural Health Product Regulations (Smith et al. 2014). Regulations for herbal medicines were introduced in Canada in 2003 as a separate law within the Food and Drugs Act (WHO 2005). Any substance administered for medical or therapeutic purposes in Canada is regarded as a medicine and is basically subject to the same system of regulatory control and requirements as set out in the Food and Drugs Act and Regulations. These are reviewed by the drug regulatory authorities before approval is given for marketing them. This is in contrast to the United States, and a large number of herbal medicines with claims of indications are, therefore, legal in Canada. The herbal substance could be a well-established drug from herbal sources such as Vincristine and Vinblastine, it could be a plant juice like an extract from hawthorn berries, or it could be an herbal tea containing chamomile. NHPs are defined as vitamin and mineral supplements, herbal and plant-based remedies, traditional medicines like traditional Chinese medicines (TCMs) or Ayurvedic (Indian) medicines, omega 3 and essential fatty acids, probiotics, and homeopathic medicines (Fan et al. 2012). All these medicines require a drug identification number (WHO 1998), just like allopathic drugs, and the same system is followed for applying for a number and review of application. For a subcategory of herbal medicine (known as traditional herbal remedies), the supporting data required by the Canadian regulatory authorities were less stringent versus that for conventional drugs. It was expected that the traditional herbal remedies will be used for minor, self-limiting conditions. The act also does not distinguish drugs on the basis of origin (synthetic and natural) but rather on the basis of purpose for which the substance is manufactured or sold. The reasoning is that the herbal remedy used in such conditions is generally less harmful and therefore safe. The actual toxic effect associated with the use of a particular plant, in fact, has no relationship with the indication of the use of that plant. The author opined that herbal medicine to be used for treating mild allergy could induce more side effects than an herbal medicine that is used for the treatment of cancer. An expert committee on herbs and botanicals was established in 1984 to review the safety and regulatory control mechanisms for herbal products that are designated as foods. Those unacceptable were added to a hazardous list in the Drugs and Food Act and Regulations and were unacceptable for sale (Chalut 1999). Various monographs are being referred for herbal medicines. GMP rules apply for manufacturing. Over 10,000 herbal medicines were registered according to a WHO survey. A PMS system is in place (WHO 2005). The regulatory control for NHPs is excellent in Canada.

5.21 CENTRAL AFRICAN REPUBLIC

Though an expert committee on TM/CAM was formulated in 1995 and a national office was established in 1997, a national policy, and laws and regulations are yet to be developed in the country. There are no research institutes, registration system, and manufacturing and safety assessment regulatory requirements in Central African Republic. National pharmacopoeias or monographs and PMS systems are being developed. No restrictions exist on the sale of herbal medicines (WHO 2005).

5.22 CHAD

The National Research Institutes on Herbal Medicines was founded in 1993, and a national office (under the Ministry of Health) was established in 2001, but the current regulatory scenario is dismal. There are no regulations for herbal medicines, and no claims can be made under law. There are no national pharmacopoeia, monograph, safety assessment, or manufacturing guidelines and no registration system in place. There are no restrictions on the sale of herbal drugs. A PMS system is being developed (WHO 2005).

5.23 CHINA

The most important agencies in China for the regulation of traditional medicine are the State Food and Drug Administration (SFDA) and the State Administration of Traditional Chinese Medicine. Traditional Chinese medicines are regulated as drugs, and the same laws and rules apply for conventional drugs (Abbott 2014). National regulations on herbal medicines have been issued, and *Chinese Pharmacopeia* (materia medica) was published in 1963. The pharmacopeia contains 990 monographs. The same GMP rules apply for manufacturing of herbal medicines as for conventional drugs (WHO 2005). Approval requires submission of preclinical and clinical data in addition to certain TCM-specific regulations such as adherence to pharmacopeias and/or monographs. Those herbal products that are ingredients in food without health claims do not need registration and are governed by general food regulations. Herbal products are categorized as traditional Chinese medicines, which refer to medicinal substances and their preparations that are used under the guidance of traditional Chinese medical theory, whereas natural medicinal products refer to natural medicinal substances, and their preparations are used under the guidance of modern medical theory. Both must follow the national drug standards. The former standard is defined by the *Pharmacopoeia of People's Republic of China* and specifications that are approved by the SFDA (Fan et al. 2012). The standards for the herbal granules (extracts for single herbs or combination classical formulae) will be available in the 2015 edition. Fan et al. (2012) discussed recently the various issues that are related to future development of global regulations of Chinese herbal products. The regulatory situation in China is quite different from other countries as there were 102 TCM products in the national essential drugs list, and these products are recommended to treat chronic diseases. In the national health insurance drug list, there are 683 TCM products. The TCM industry, thus, is 30% of the total value of the pharmaceutical industry in China (Pelkonen et al. 2012).

5.24 COMOROS

The National Research Institute on Traditional Medicine was founded in 1979, and a national office (under the Ministry of Health) was established in 2002 in the Union of Comoros. There are no regulations on herbal drugs; no claims can be made legally; and there is no national pharmacopoeia or monograph, no manufacturing or safety assessment requirements, no registration system, and no postmarketing system in place. Herbal medicines are sold freely without any restrictions (WHO 2005).

5.25 CONGO

There is no national policy on TM/CAM. No regulatory status for herbal medicines exists, and no claims may be made under law. A national pharmacopoeia is currently being developed. The *African Pharmacopoeia* (1985) is currently used. No national monograph or registration system exists in the country. The PMS system is being developed. There are no restrictions on the sale of herbal medicines. The regulatory requirements for safety are limited to special requirements that include traditional use without harmful effects and documented research on similar products. No control mechanism exists for these regulations (WHO 2005).

5.26 CZECH REPUBLIC

National laws and regulations on herbal medicines were introduced in 1997 by the same law that regulates conventional drugs. Claims may be made about herbal medicines, which were sold as prescription or OTC medicine. The *Cesky lekopis* 2002 (*Pharmacopoeia bohemica MMII*) is the national pharmacopoeia containing two legally binding parts: the *European Pharmacopoeia* in translation and national monographs. Regulatory requirements for manufacturing included adherence to pharmacopoeia, monographs, and GMP rules that are used for conventional pharmaceuticals.

Safety requirements were the same as conventional pharmaceuticals, and assessment during the registration process is guaranteed. There were 230 registered herbal medicines up to November 2003. A national PMS system exists (WHO 2005).

5.27 DENMARK

Regulations on herbal medicines began in 1992. There was no national pharmacopoeia, and hence *European Pharmacopoeia* was used (WHO 2005). Natural remedies are governed by the Danish Ministry of Health Regulations as follows. (1) medicinal products in which the active substance or content is exclusively composed of naturally occurring substances in quantities that are not greater than those that are found in nature. (2) These apply to natural remedies that are intended for oral use and local application on the skin or mucus membranes. It does not apply to prescription drugs and homeopathic medicines. (3) Combinations of vitamins and minerals cannot be marketed as natural remedies. (4) Proof of quality, safety, and efficacy must be given (WHO 1998). A bibliographic document must accompany the application for approval supporting its therapeutics claims. The scientific literature cited should be from publications from Europe or North America (WHO 1998). PMS exists, and there were 170 registered herbal medicines according to the WHO (2005).

5.28 EQUATORIAL GUINEA

Regulations of herbal medicines were established in 1985 through a special legislation applying only to herbal drugs. By law, claims may be made about herbal drugs that are sold as OTC medicines. There is no national pharmacopoeia. National monographs exist in the *Recetario Plantas Medicinales de Equatorial Guinea* (1996), which contains 18 monographs. The information is not legally binding. No requirements exist for manufacturing or for registration, and there is no PMS system or restrictions on sale (WHO 2005).

5.29 ESTONIA

The Medicinal Products Act 1996 lays down general requirements and procedures for approval and registration (WHO 1998). Application with documents concerning chemical, pharmacological, toxicological, and clinical information should be submitted with a summary of product characteristic and information on the price and design of the package (WHO 1998). Regulations on herbal drugs began in 1998 and are sold as prescription, OTC drugs, or as dietary supplements. The *European Pharmacopoeia* and national monographs are used and are legally binding. The *British Herbal Pharmacopoeia* and WHO monographs are also used. Regulatory requirements include adherence to pharmacopoeia for manufacturing and following the same GMP rules as conventional drugs. Safety requirements include special requirements of traditional use without demonstrated harmful effects, reference to documented scientific research on similar products, or may include requirements that are similar to other pharmaceuticals. According to a

WHO survey, 85 herbal medicines were registered, and a PMS system was in place (WHO 2005).

5.30 ETHIOPIA

The national policy on TM/CAM was formulated, and regulations or laws were introduced in the Federal Democratic Republic of Ethiopia in 1999. The Drug Administration and Control Authority (DACA) of the Ministry of Health governs these issues. The Drug Research Department of the Ethiopian Health and Nutrition Institute and the DACA expert committee cover all aspects of traditional and herbal medicines. Herbal medicines are sold with medical claims, but no regulatory status exists for these drugs. No national pharmacopoeia or monograph, regulatory requirements for manufacturing or safety assessment, registration system, or PMS system existed. There were no restrictions on the sale of herbal medicines in Ethiopia (WHO 2005).

5.31 FINLAND

Administrative Regulation 9/93 regulates the status of herbal products (WHO 1998). These can normally be sold as health products in shops, department stores, and pharmacies if not registered as medicines. For the latter category, an application has to be submitted to the National Agency for Medicines for approval. A full dossier on quality, safety, and efficacy is mandatory in accordance with the EU directives and EP. The documents should include full information on raw materials; traditional use; ingredients with methods of identification of active ingredient(s); chemical characteristics; pharmacological, toxicological, and pharmaceutical studies; forms of usage; range of doses; indications; contraindications; side effects; interactions with conventional medicines; GMP by manufacturers; labeling, and so forth. Bibliographies of scientific literature verifying the safety and efficacy of the product must also be submitted with the application for authorization (WHO 1998).

5.32 FRANCE

The French Republic does not have a national policy, laws or regulations, national office, expert committee, or national research institute on TM/CAM. Regulation of herbal medicine in France began in 1985. The same laws and regulations are used for herbal medicines and conventional pharmaceuticals. Herbal products were regulated as OTC medicines, and medical claims could be made about them. No national pharmacopoeia or monographs existed. The same GMP rules are required for the manufacture of herbal medicines as for conventional pharmaceuticals. Compliance with these requirements was ensured through inspections. Special requirements for traditional use without demonstrated harmful effects has been included for herbal remedies. In France, 787 herbal medicines were registered; none was listed on the national essential drugs list. A PMS system that includes pharmacovigilance for herbal medicines has been well established (WHO 2005). This is proven by the example of 30 cases of hepatitis as ADRs reported for germander following, which it was withdrawn in the French market (Castot et al. 1997).

5.33 GEORGIA

Regulation of herbal medicines was introduced in 2002 with laws and regulations similar to those that regulate conventional pharmaceuticals. A national pharmacopoeia was published in two volumes in 2000 and 2003, and is legally binding. National monographs on herbal medicines were under development. Manufacturing regulatory requirements for herbal medicines are adherence to pharmacopoeias and monographs. Safety requirements for herbal medicines are the same as those for conventional pharmaceuticals. No control mechanism exists to ensure compliance. There were 181 registered herbal medicines according to a WHO survey, but none were included on a national essential drug list. No PMS system for herbal medicines existed (WHO 2005).

5.34 GERMANY

The national laws and regulations in Germany were introduced in 1976 and have been updated and amended from time to time. Herbal medicines are sold as OTC medicines. The national pharmacopoeia, the *Deutsches Arzneibuch* (*German Pharmacopoeia*) and the *European Pharmacopoeia*, are legally binding. Compliance with the pharmacopoeia was ensured through regular inspections. Safety regulatory requirements include those that are required for conventional pharmaceuticals, and implementation was ensured through pharmacovigilance (WHO 2005). In Germany, herbal drugs, which are sold as *phytomedicines*, are subjected to the same criteria for their safety, efficacy, and quality as applicable to other drugs. Regulatory controls are, therefore, considered necessary to safeguard drug interactions with herbal drugs (George 2011). There are approximately 3500 herbal medicines that are registered in Germany. The PMS system was established in 1978 and included monitoring for adverse effects of herbal medicines (WHO 2005).

5.35 GHANA

A national research institute was established in 1975, an expert committee and national office was put into place in 1999, and a national policy on TM/CAM was formulated in 2002, in the Republic of Ghana. Herbal regulations began in 1992 through Food and Drugs Law, which also regulates conventional pharmaceuticals. The *Ghana Herbal Pharmacopoeia* was published in 1992 but is not legally binding. Regulatory requirements for manufacture followed the same GMP rules as for conventional medicines. Safety assessment requirements were ensured through a pharmacovigilance center. There were 340 registered herbal medicines according to a WHO survey but none on the WHO National Essential Drug List. A PMS system has existed for herbal medicines since 2000 (WHO 2005).

5.36 GREECE

The Ministry of Health Regulation 1994 governs herbal medicines. It applies to such products that contain only plants or plant preparations as active ingredients. The regulation does not apply to products that are used in food or beverages unless claims or advertisements are made for their therapeutic indications as medicines. If this is done, the plant product must fulfill all the requirements of the law. Application with a full dossier, which complies with EU legislation, must be submitted to the National Pharmaceutical Office, which will grant or reject the registration. Detailed instructions for the content of the package insert and label are mandatory (WHO 1998).

5.37 GUINEA

The Division de Medicine Traditionalle (Traditional Medicine Division) under the Ministry of Health began in 1977. A national policy was formulated in 1994, followed by the introduction of laws and regulations and expert committee on TM/CAM in 1999 in the Republic of Guinea. A National Institute on Herbal Medicine was established in 2001. Regulations and laws on herbal medicines in the country were partly the same as those for conventional pharmaceuticals. No national pharmacopoeia exists; other pharmacopoeias are used but are not legally binding. National monographs exist in the *Plantes Medicinales Guineennees* (1997), but they are not legally binding. Regulatory requirements for manufacturing include adherence to the information that is contained in the pharmacopoeia and monographs and special GMP for herbal medicines. The regulatory requirements for safety assessment include the special requirements of traditional use without demonstrated harmful effects and reference to documented scientific research on similar products. There is no control mechanism for the implementation of these requirements (WHO 2005).

5.38 HUNGARY

The laws and regulations on naturopathic activities in Hungary were issued in 1987, and a national policy was formulated on TM/CAM in 1997. Herbal medicinal products may be sold as traditional herbal products called "healing products of paramedicine" (having therapeutic effects but not considered as medicaments) or as herald medicines, which are considered to be conventional pharmaceutical products. The regulation for traditional herbal products was issued in 1987 according to which the traditional herbal product may be approved if its composition and components are known, the quality of the components and product is determined and constantly ensured, its safety in the doses to be administered is proven, and the conditions of its production meet the public health regulations. The regulations for herbal medicines were placed into law in 1988, and regulations in 2000 and 2001 that included some special quality requirements for herbal medicines. The GMP rules used for conventional pharmaceuticals are also required for the manufacture of herbal medicine (WHO 2005). The quality should conform to *Hungarian Pharmacopoeia* and *European Pharmacopoeia*. The safety and efficacy of herbal medicinal products may be proved using the same requirements as those for conventional pharmaceuticals, including preclinical and clinical trials. A PMS system including adverse effect monitoring was established in 1970 (WHO 2005).

Though an extraordinary diversity has been found in laws and regulations of CAM within the EU, a harmonization of laws is underway in many countries. In addition to harmonization with the EU, the modifications of the decree regulating CAM in Hungary are being prepared following professional recommendations. The outdated decree will be reconsidered, as 17 years of practice have shown which practices are worthy to be maintained, taught, and applied within the healthcare system and which should be omitted (Gabriella et al. 2013).

5.39 ICELAND

The Republic of Iceland does not have a national policy, laws and regulations, national program, national office, or national research institute on TM/CAM. Regulation of herbal medicine in Iceland began in 1997. An expert committee on TM/CAM was established in 2000 as part of the Icelandic Medicine Control Agency. No national pharmacopoeia or monographs exist; the *European Pharmacopoeia* is used and is considered to be legally binding. The same GMP rules apply as for conventional pharmaceuticals. Safety regulatory requirements include traditional use without demonstrated harmful effects and reference to documented scientific research on similar products, provided by manufacturers. Eleven herbal medicines were registered in Iceland according to a WHO survey; none was included on a national essential drugs list. The PMS system used is similar to that for conventional pharmaceuticals. It included adverse effect monitoring (WHO 2005).

5.40 INDIA

Medicinal plants are used for therapeutic purposes in India in several ways. The country has a well-developed system of modern medicines. It has many plant-derived drugs (e.g., digitalis, morphine, atropine, cinchona, Vinblastine, etc.), which are regulated in the same way as for synthetic allopathic medicine, which has to be supported by full pharmacological, toxicological, and later pharmacokinetic and early clinical trial data. Several plants are used (alone or in combination with other herbal ingredients) in traditional systems of medicine (Ayurveda, Siddha, and Unani-Tibb) that are practiced in India. The regulation states that if these medicines are prepared in exactly the same manner as described in ancient texts of Indian medicine and preserved in the same way, it should be considered safe and approved for registration for release into the market. If, however, an old medicinal plant or herbal product is prepared in a different manner or contains a new ingredient (or has one ingredient less) or if it is to be used for a new indication, it is treated as a new drug entity. Application for its introduction or marketing of such products needs full supporting data as for other conventional drugs that were mentioned previously in this chapter (GCP Guidelines 2001).

The Drugs and Cosmetics Act and Rules (1940) and the Drugs and Cosmetics Act and Rules (1945) with many subsequent amendments (1949, 1950, 1951, 1955, 1960, 1964, 1972, 1982, 1995, 2005, and 2010) and the Drugs and Magic Remedies Act (1954) are applicable for the purpose. Mukherjee et al. (2007) reviewed the status and progress in regulation and control of medicinal and aromatic plants that are used in the Indian Systems of Medicine. Various aspects including manufacturing premises, heavy metal content, GMP, infrastructure development, National Medicinal

Plants Board, *Indian Pharmacopoeia*, Traditional Digital Knowledge Library, Indian Traditional Development Bill 2005, and Traditional Medicine Act 2006 were discussed. The authors tabulated the sections of the Indian Drugs and Cosmetics Act for misbranded, adulterated, and spurious drugs; manufacturing; sales; government inspectors or analysts; penalties and confiscation, and so forth. The country has two multivolume national pharmacopoeias: the *Ayurvedic Pharmacopoeia of India* and the *Unani Pharmacopoeia of India*. Both are legally binding. There were 4246 registered herbal medicines according to a WHO survey, and there is a separate essential drugs list for Ayurveda, Unani, and Siddha (WHO 2005).

The Central Drugs Standard Control Organization (CDSCO) published guidelines for clinical trials of herbal remedies and medicinal plants (GCP Guidelines 2001). The herbal products were categorized into three categories:

1. Sufficient information is known about the use of plant or its extract in ancient Ayurvedic, Unani, or Siddha literature, and the herbal remedy is being regularly used by physicians of these traditional systems of medicine for a number of years. Further, the substance is to be clinically evaluated for the same indication for which it is being used and has been described in ancient texts.
2. When an extract of a plant or isolated principle is to be clinically evaluated for a therapeutic effect, not originally described in traditional texts, or method of preparation is different, it has to be treated as a new chemical entity (NCE), and the same type of acute, sub-acute, and chronic toxicity data will have to be generated as required by the regulatory authorities before it is cleared for clinical evaluation.
3. An extract or a compound isolated from a plant that has never been in use before and has not been mentioned in ancient literature should be treated as a new drug and, therefore, should undergo all regulatory requirements for clinical trials including GMP norms, preclinical and clinical Phase I and Phase II studies, and postmarketing surveillance.

India joined the WHO ADR monitoring program in 1997, and a national pharmacovigilance program started in 2004. In July 2010, CDSCO, under the aegis of Ministry of Health and Family Welfare, Government of India in collaboration with the Department of Pharmacology, All India Institute of Medical Sciences, New Delhi, launched a pharmacovigilance program in India for protecting health of the patients by assuring drug safety (drug, biological, vaccine, diagnostics, and medical devices). It aims to monitor ADRs and create general awareness on the subject in healthcare professionals. There is a national program by the Department of Ayurveda, Yoga, Unani, Siddha, and Homeopathy (AYUSH) for the pharmacovigilance of Ayurveda, Siddha, and Unani medicine at Jamnagar, Gujarat, since 2008. This was set up in collaboration with the WHO country office for India at New Delhi.

There are multiple authorities: (1) the Drug Controller General of India, Department of Health, Ministry of Health and Family Welfare, Government of India, New Delhi, and (2) the Secretary, Department of AYUSH, Ministry of Health and Family Welfare, Government of India, New Delhi, for registration, distribution, and marketing of these drugs in the country. The Indian Council of Medical Research (ICMR) has a Toxicology Review Panel (TRP) to advise DCGI or AYUSH to permit (or reject) clinical trials of herbal drugs in human subjects. In response to our request for information on the subject, ICMR searched data from its archives for the last 10 years. It revealed that TRP did not recommend some proposals for clinical trials in human subjects due to the following reasons: (1) The phase 1 data submitted was not considered adequate for phase 2 clinical trial, (2) the detailed composition of the herbal formulation was not provided, (3) the quality control and stability studies were not performed, or (4) a proper study design was lacking. Information on names of herbal drugs or formulations or the manufacturer seeking permission for clinical trials in human subjects was considered confidential (and therefore not revealed); the proposals that were not approved were related to clinical trials in three areas: (1) contraception, (2) diabetes mellitus type II, and (3) cancer (Katoch and Vijay Kumar 2012).

Heavy metals are sometimes the active ingredients of Indian traditional medicine rather than a contaminant (Abbott 2014). The Government of India has made heavy metal testing mandatory after the publication of Saper et al. (2004). Heavy metals may not be present above the permissible limit, and labeling must indicate the presence of heavy metals (Abbott 2014). Evidence-based studies are becoming interestingly essential for establishing the safety and efficacy of herbal products for domestic sale and export. Regulatory harmonization becomes essential to mitigate delays in commercialization across countries. The Indian industry perspective was recently discussed (Sahoo and Manchikanti 2013).

5.41 INDONESIA

Laws and regulations on TM/CAM were introduced in 1993. National research institutes on traditional and herbal medicines were established in 1976. The Minister of Health issued a decree in the same year to control the import of crude drugs that are marketed in Indonesia. The importer should be registered with the Ministry of Health and should fulfill many conditions for registration of the herbal product according to the *Indonesian Pharmacopoeia*. The Directorate General of Drug and Food Control has also published six volumes of the *Materia Medica* Indonesia. It contains 350 crude drug monographs and is used as a formal quality requirement book for crude drugs in Indonesia. The national policy on the development of traditional medicine was formulated in 2000 followed by the establishment of a national office under the National Agency for Food Control in 2001 and the launching of a national program on TM/CAM in 2003. Adherence to *Farmakope Indonesia* and *Materia Medica Indonesia* is legally binding. Special GMP rules apply for the manufacture of herbal medicines. Safety requirements include traditional use without documented harmful effects, reference to documented scientific research on similar products, toxicity data, and laboratory testing. PMS is in place. There were 8632 registered herbal medicines (WHO 1998, 2005).

5.42 IRAN

In Iran, the specific regulations for registration of herbal products were documented by the National Expert Committee in 1996, and regulation of all herbal medicines is under the authority of the pharmaceutical department of the Ministry of Health (http://www.itmrc.org/regulator1.htm). There is a national formulary containing 70 monographs but it is not legally binding (WHO 2005). A PMS system exists.

5.43 IRELAND

The Guidelines for Application for Authorization of Herbal Products were issued by the National Drug Advisory Board in 1985. This board was replaced by the Irish Medicines Board in 1996. The guidelines contained detailed requirements for the plant raw material and for the finished product. These have, largely, been superseded by the European note for the quality and safety of herbal remedies. Safety and efficacy requirements are now consistent with those for conventional pharmaceutical products (WHO 1998). The *European Pharmacopoeia* is used in lieu of a national pharmacopoeia and is legally binding. Regulatory requirements for manufacture include adherence to pharmacopoeias, monographs, and the same GMP rules as for conventional drugs. There is no registration system. The PMS system is the same as for conventional drugs (WHO 2005).

The European Traditional Herbal Medicinal Products Directive (2004/24/EC) went into effect in Ireland in 2007.

5.44 ISRAEL

There is currently no specific regulation of herbal medicines in Israel. Herbal medicines may be considered as dietary supplements. No national pharmacopoeia exists; instead, the *Homeopathic Pharmacopoeia*, *British Pharmacopoeia*, *French Pharmacopoeia*, and the *United States Pharmacopoeia* are used. The regulatory requirements for the manufacture of herbal medicines is the same as for conventional pharmaceuticals. For safety, herbal medicines are screened to restrict those that are not suitable for food use. According to a WHO survey, no registration system existed, and there were no restrictions on the sale of herbal medicines, but a PMS system existed (WHO 2005).

5.45 JAPAN

Traditional medicine in Japan is divided into Kampo and non-Kampo crude herbal products. Kampo is a Japanese traditional medicine that has its roots in China. Japanese practitioners gradually refined and improved and validated the original formulae of Chinese through clinical experience. Kampo medicines are accepted by the Japanese health insurance system. The quality of these herbs is regulated by determining the main chemical constituents according to the guidelines in the *Japanese Pharmacopoeia*, which lists both crude drugs and Kampo extracts. While most Japanese doctors and patients are comfortable using them, the efficacy and toxicity of Kampo medicines are still uncertain because the double-blind placebo-controlled clinical trials have been rare (Sensho 2014). In the early 1970s, the "210 OTC Kampo Formulae" was published by the Ministry of Health and Welfare. These were revised in 2008, and now, 294 Kampo formulae are listed. The GMP Guidelines for ethical Kampo products were published in 1987 and revised in 2012. Both approval and quality standards play an important role in the regulation of Kampo products (Maegawa et al. 2014).

5.46 KENYA

According to a WHO survey, herbal medicines were not regulated in the Republic of Kenya. The country has no pharmacopoeia or monographs. National policy and laws and regulations on TM/CAM are being developed. Special regulatory requirements for the safety assessment of traditional use without demonstrated harmful effects and reference to documented scientific research on similar product requirements apply to herbal medicines. These have been established by the Kenya Medical Research Institute, but no control mechanism exists to ensure their implementation. Herbal medicines in Kenya are sold without restriction (WHO 2005).

5.47 KOREA

In South Korea the Korea Food and Drug Administration (KFDA) and Pharmaceutical Affairs Act regulate both conventional and traditional medicines. There is a *Korean Herbal Pharmacopoeia* and "Regulations on Limits and Test Methods for Residues and Contaminants in Herbal Medicines." The latter restricts the amount of hazardous substances such as heavy metals, pesticides, and aflatoxins. The manufacturing of health-functional foods must be approved by the KFDA according to the Health Functional Food Act (Fan et al. 2012). A PMS system is in place (WHO 2005).

5.48 MALDIVES

Though a medicinal policy in the Republic of Maldives was formulated in 1999, the country does not regulate herbal products. These are classified as OTC medicines for self-medication only. No claims may be made by law. There are no regulatory requirements for manufacturers, no national pharmacopoeia, or monographs. Safety requirements are limited to documented scientific research on similar products (WHO 2005).

5.49 MALI

A division of traditional medicine in Mali, which was also a collaborating center of WHO, started in 1974 and carried out activities such as survey of practitioners; identification of medicinal plants in Mali and their botanical, chemical, and pharmacological studies; development of improved traditional medicines; quality control; and training programs (WHO 1998). The *African Pharmacopoeia* is used but is not legally binding. A national monograph exists in the *Formulaire Therapeutique* (1998) but is not legally binding. Since 1990, improved traditional medicine called Médicaments Traditionnels Améliorés (MTA) have been included on the Essential Drugs List of Mali, are included in the Malian National Formulary alongside conventional drugs. For marketing authorization, a dossier of information on safety and efficacy is submitted (Willcox et al. 2012). Regulatory requirements for manufacturing include adherence to information in pharmacopeia or monographs as well as good conditions for harvest, drying, extraction, packaging, and quality control. Implementation is ensured by laboratory testing for content and foreign substances and by ensuring correct identification of plants and standardized extracts. Safety assessment implementation is ensured through toxicity testing. The country has a registration system and PMS system (WHO 2005). In Mali, traditional medicine is integrated into the health system, and traditional medicine is prescribed under the national health system (Fan et al. 2012).

5.50 MAURITIUS

A survey was carried out in the early 1990s, and a study was initiated on the medicinal and aromatic plants of States of Indian Ocean by the EU under the aegis of the Indian Ocean Commission. The results were good. More than 600 plants entering the pharmacopoeia were identified. Phytochemical, botanical, and bibliographic information were made available. The physiochemical properties of some medicinal plants, pharmacological properties of some plant extracts, and measures for the control plant material to guarantee the safe and correct use of herbal drugs were established (WHO 1998).

5.51 MYANMAR

In Myanmar, the Traditional Medicine Drug Law was enacted in 1996 to ensure the quality, safety, and efficacy of traditional medicines. No national pharmacopoeia, essential drugs list, or PMS system exists. The *Monograph of Myanmar Medicinal Plants* was published in 2000. The regulatory requirements for herbal medicines are limited to special GMP rules. Safety requirements include traditional use without demonstrated harmful effects and reference to documented scientific research on similar products. These requirements are enforced through inspection, laboratory analysis for safety, and market surveys. There were 3678 registered traditional medicines in Myanmar according to a WHO survey (WHO 2005).

5.52 NEPAL

The Department of Drug Administration is the regulatory authority in Nepal (http://www.dda.gov.np). The Drug Act 1978 regulates herbal and conventional medicines in Nepal. Manufacturing requirements include the same GMP rules as for other medicines. A PMS system is in place (WHO 2005).

5.53 NETHERLANDS

The Netherlands does not have a national policy, laws or regulations, or national research institutes on TM/CAM. Herbal medicines are regulated under the same laws as conventional pharmaceuticals.

The *European Pharmacopoeia* is used and is considered legally binding. The GMP rules that are used for conventional pharmaceuticals apply to herbal medicines. Safety requirements are the same as for conventional pharmaceuticals (WHO 2005).

5.54 NIGERIA

Herbal medicines are regulated as dietary supplements, health foods, functional foods, and as an independent regulatory category (WHO 2005). The National Agency for Food and Drug Administration and Control (NAFDAC), established in 1993, is the regulatory authority in Nigeria. It developed guidelines for the registration and control of herbal medicinal products and related substances. According to these, herbal medicinal products manufactured on a large scale must be registered and approved by the NAFDAC. PMS is mandatory (Osuide 2002). A national pharmacopoeia is available.

5.55 NORWAY

Herbal products in Norway require registration with the Norwegian Medicines Control Authority. Guidelines for such drugs, published in 1994, were largely based on European directives. These were later simplified (WHO 1998). A national policy on TM/CAM was formulated in 2002 followed by national regulations and laws in 2004. Prescription herbal products are governed by European guidelines for quality, safety, and efficacy according to monographs in EP, GMP, and WHO guidelines 1991. The latter are less stringent for herbal remedies that are intended to treat minor ailments and history of long traditional use. Four categories for documentation were listed for such drugs. Herbal teas should not contain more than 70% by weight of active ingredient(s). Norwegian guidelines contain requirements for mandatory information on labels. There are no national pharmacopoeia or monographs. The *European Pharmacopoeia* and monographs are legally binding (WHO 1998, 2005).

5.56 POLAND

The registration and marketing of medicinal products are governed by the Pharmaceutical Act Law of 2001, and their distribution is regulated by Good Distribution Practice Ordinance 2002, 2009, Minister of Health. Some other laws and rules, for example, the Pharmaceutical Act 2004, Freedom of Business Activity Act 2007, and Ordinance 2008 regulate marketing authorization at various levels. Czerw and Bilinska (2013) stressed that the legal regulations governing these products is of vital importance. Their level of detail and severity are necessary to ensure the safety of patients and honest competition.

As a member state, Poland can have a simplified registration of THMPD in the EU.

5.57 PORTUGAL

The regulations or laws for herbal medicines are similar to those for conventional pharmaceuticals. Herbal medicines are regulated as prescription and OTC medicines.

Medical claims may be made about herbal medicines. There is a national pharmacopoeia, *Farmacopoeia Portuguesa*, but no national monograph exists. The same GMP rules used for conventional pharmaceuticals are used for herbal medicines, although a control mechanism exists for these requirements. There is no regulation system or essential drugs list for herbal medicines. The PMS system, which includes adverse effect monitoring, was established in 1995 in Portugal. Herbal medicines are sold in pharmacies as prescription and OTC medicines (WHO 2005).

5.58 ROMANIA

Regulation of herbal medicines was introduced in 2002 in the same law governing the regulation of conventional pharmaceuticals. *Romanian Pharmacopoeia*, published in 1993, contains 85 monographs and is legally binding. The manufacturing and safety requirement for herbal medicines are the same as those that are for conventional pharmaceuticals. Monitoring the safety of the products on the market is ensured through a pharmacovigilance network (WHO 2005).

5.59 RUSSIA

In Russia, a federal law about the "circulation of drugs" regulates pharmaceutical preparations including herbal medicine preparations (HMPs), as officially known. HMPs are classified into medicinal plant materials, galenic formulations, active pharmaceutical ingredients, and combined phytopreparations. HMPs must follow the national drug standards and the *12th State Pharmacopoeia of Russian Federation*. More than 600 HMPs have been registered for medication (Shikov et al. 2012). The same GMP rules apply as those that are required for conventional pharmaceuticals. The requirements for the assessment of safety of herbal medicines are the same as for conventional pharmaceuticals with additional requirements, namely, radioactivity control. A PMS system is in place (WHO 2005).

5.60 SOUTH AFRICA

The protection of the intellectual properties of traditional healers has become an increasingly important aspect of the regulatory scenario (Fan et al. 2012). There are about 200,000 traditional healers in the country (WHO 1998). The Traditional Medicine Programme at the Development of Pharmacology, University of Cape Town, participated in formulating an outline proposal for the registration and control of traditional medicines in 1994 (WHO 1998). Though South Africa has national laws that are related to traditional medicine, they are not applied in practice (Fan et al. 2012).

5.61 SPAIN

The Spanish Medical Products Act No. 25 of 1990 regulates both herbal medicines and conventional pharmaceuticals. The *Royal Spanish Pharmacopoeia* was published in 2003. A national monograph also exists. Both are considered legally binding. Regulatory requirements for the manufacture of herbal medicines include adherence to pharmacopoeia and monographs, the same GMP rules as for conventional pharmaceuticals, and special GMP rules. Safety requirements include those that are used for conventional pharmaceuticals as well as a special requirement of traditional use without demonstrated harmful effects and reference to documented scientific research on similar products. If the product has a history of traditional safe use, the requirements are less strict. Compliance with requirement is ensured through the national pharmacovigilance system. A PMS system was established in 1985 (WHO 2005). Herbal products are categorized as "real medicinal products" with all marketing authorization requirements and traditional "herbal products" (HPs) for which rules are less stringent. These HPs are further categorized depending upon practical use or quality or efficacy standards. Peschel (2007) opined that Spain will face a radical change by EU legislation (2004/24/EC), and a major issue for reclassification is that HPs, which are considered as medicine, can be distributed only via pharmacies.

5.62 SRI LANKA

In the Democratic Socialist Republic of Sri Lanka, the laws and regulations on TM/CAM were issued in 1961 and a national program in 1982. The Department of Ayurveda in the Ministry of Health was established in 1961 and a national research institute on traditional medicines, complementary medicines, and herbal medicines in 1962. Herbal medicines do not have any regulatory status; they are sold with medical, health nutrient contents, and structure or function claims. The national pharmacopoeia, the *Ayurveda Pharmacopoeia*, was published in 1979 and a compendium of medical plants containing 100 national monographs in 2002. The information contained therein is legally binding. Regulatory requirements for manufacturing include the same GMP rules that apply to conventional pharmaceuticals. There are no safety requirements. There was no national registration system and PMS according a WHO survey (WHO 2005).

5.63 SWEDEN

The Medicinal Products Agency issued guidelines for authorization to market natural remedies in 1994. These include (1) finished products that are intended for administration in human beings or animals for prevention, diagnosis, relief, or cure of diseases or symptoms of diseases; (2) a medicine in which the active ingredient or ingredients are derived from natural sources, for example, plants, animals, bacterial culture, minerals, or salt solution. The chemical, pharmaceutical, safety, and efficacy document is assessed by the agency. Manufacturing requirements include adherence to the *European Pharmacopeia*. A PMS system is available. Natural remedies are sold freely by anyone in Sweden (WHO 1998, 2005).

5.64 SWITZERLAND

The Swiss Agency for Therapeutic Products (Swissmedic), established in 2002, is the national office for TM/CAM. Herbal medicines were authorized as medicinal products even before the federal law on medicinal products and medical devices, which regulates conventional medicine, was adopted in 2000. Simplified marketing authorization for some TM/CAM may apply but only after documented quality, efficacy, and safety requirements. The *Swiss Pharmacopoeia* (containing monographs on herbal drugs) and *European Pharmacopeia* are used. Manufacturing requirements are similar to conventional medicines. PMS is in place (WHO 2005).

5.65 TAJIKISTAN

According to a WHO survey, a national policy and a national program on TM/CAM were in development in Tajikistan. Regulation of herbal medicines was introduced in 2001, comprising the same laws and regulations as for conventional pharmaceuticals. Herbal medicines have no regulatory status. There was no national pharmacopoeia; the *State Pharmacopoeia of the USSR* was used. The manufacturing regulatory requirement in Tajikistan is limited to adhere to information in pharmacopoeia and monographs; no specific details of the structure of the control mechanism were available. A PMS system was being developed (WHO 2005).

5.66 THAILAND

A national policy and program on herbal medicines were issued in 1993 (WHO 2005). The Thailand Drug Act BE 2510, established in 1967, regulates both traditional and modern drugs. The drug act distinguishes between "household medicine" and manufactured formula herbal drugs; while the former may be used by any person or a traditional medicine practitioner for his patients, the latter needs to be registered, with names and quantities of the ingredients on the label or leaflet of the products before permission for distribution or marketing is accorded (WHO 1998). The Thai FDA classifies herbal medicines into four categories: (1) traditional drugs (based on traditional knowledge and vast experience), (2) modified traditional drugs (where dosage forms of [1]) has been modified into modern forms), (3) modern herbal medicines or phytopharmaceuticals (composed of active plant materials that are derived from scientific research and are classified as modern medicines), and (4) new drugs (from herbs that are developed and are purified isolated active substances of which chemical structures are identified as new chemical entities). The new regulations will be according to ASEAN guidelines, and to harmonize the same, the Thai FDA issued Ministry of Public Health regulations on the "Submission of license and issuance of license for the production, sale or import of traditional medicines into the Kingdom BE 2555" in 2012 (Thailand Drug Act BE 2510 1967).

5.67 TURKEY

Turkey is rich in flora of medicinal plants. The country mainly relies on modern medicine for healthcare; traditional therapies only have a limited use. Prior to 1984, there were no regulations for herbal drugs and crude drugs that were sold in Akhtar shops by persons with no special training. The Ministry of Health listed plants that were sold by these shops in October 1985. This list included mainly crude herbs and their parts. Later, regulations for the sale of poisonous plants, the establishment of herbal manufacturing premises and GMP rules in 1986, and licensing in 1995 were introduced (WHO 1998). The Turkish and European pharmacopoeias are used. PMS was established in 1985 (WHO 2005).

5.68 UGANDA

Though the national pharmacopoeia entitled *A Contribution of the Traditional Medicines Pharmacopoeia of Uganda (1993)* exists, the information is not legally binding. There are no manufacturing regulatory requirements for herbal medicines, no medicines were registered though a registration system was established in 2002, and no PMS system is in place. In Uganda, herbal medicines are sold in pharmacies as OTC medicines, by peddlers, and in food markets, without restriction (WHO 2005). The "Guidelines for Regulation of Traditional/Herbal Medicines (Local) in Uganda" were issued by the National Drug Authority in 2009 incorporating requirements for the manufacture, registration, and sale of herbal medicines. According to these guidelines, herbal medicines are classified as (1) home remedies, (2) galenicals, and (3) traditional medicines. The guidelines also specify minimum requirements for efficacy and safety (Guidelines for Regulation of Traditional/Herbal medicines [Local] in Uganda 2009).

5.69 UKRAINE

The National Policy on TM/CAM in the Ukraine was established in 1992 and the regulation of herbal medicines in the Ukraine was introduced in 1992; these laws and regulations are partly the same as those that govern conventional pharmaceuticals. In lieu of a national pharmacopoeia, which is in development, the *State Pharmacopoeia of the USSR* was used according to a WHO survey. Regulatory requirements for manufacturing include adhering in pharmacopoeias and monographs and the GMP rules that are required for conventional pharmaceuticals. No specific information was available about the control mechanism that was used. Safety requirements are the same as for conventional pharmaceuticals.

There is a registration system for herbal medicines. A PMS system, which includes adverse effect monitoring, was established in 1992 (WHO 2005).

5.70 UNITED KINGDOM

The Medicines Act of the United Kingdom (1968) distinguished "licensed herbal medicines," which require marketing authorization, from "herbal remedies," which were exempted from licensing requirements. PMS was established for both categories. The *British Pharmacopoeia* is available (WHO 2005). All herbal medicines offered for sale for therapeutic purposes have to be approved by the Medicines Commission and registered by the appropriate authority. The requirement needed by the government for herbal products is quite different from those that are needed for synthetic drugs that are being introduced in the market for the first time. The Medicines Commission Secretariat (assisted by the pharmacologist and the pharmacognosist) reviews what is known about the plant, in a spirit of pragmatism and tolerance since the plants have been used in the United Kingdom for centuries, to decide whether the substance can be allowed for sale in the country. The existing knowledge about the herbal product (from within or outside the country) for varied aspects is thoroughly scrutinized for (1) evidence for its intended efficacy, (2) reported side effects, (3) expected side effects because of its structure, and (4) use in other countries and experience in those countries that was documented in scientific books or journals. Another aspect is to look at is the type of disease for which the product is to be used. A plant substance to be used for myocardial infarction would undergo much more rigid scrutiny than a plant product to be used for a backache or indigestion. The authorities are aware that the ethnic migrant groups in the United Kingdom bring with them their own herbal remedies (Ayurvedic, Unani, or Chinese) and normally do not interfere with their use so long as (1) these medicines are restricted to use by these communities with a firm faith and belief in these remedies, and (2) this latitude is not being misused by advertising and selling these products. This attitude of tolerance and tacit agreement for the use of herbal drugs that are not registered in the United Kingdom, by ethnic groups from India, Pakistan, and Bangladesh has, and by large, been successful for more than four decades. Recently, a new regulation was introduced (Human Medicine Regulations 2012). It has superseded the Medicine Act of 1968. Now, products sold OTC must be authorized and have a Traditional Herbal Medicinal Products Directive (THMPD) license that is covered by part of this regulation. A recent note formally ends transitional arrangements or a grace period that allowed continued sale of unlicensed herbal medicinal products (under Directive 2004/24 EC) from May 1, 2014 (MHRA 2013). Following its implementation, all manufactured herbal medicines will need full marketing authorization for their lawful sale. Herbal retailers will no longer be able to sell unlicensed herbal medicines or finished products that are not registered under the Traditional Herbal Medicine Registration Scheme. Qualified medical herbalists will only be allowed to provide preparations after a one-to-one consultation made from a basic herbalist's stock at the practitioner's premises, where the general public can be excluded. Eventually, a clear-cut application of well-formulated regulations to address the safety concerns for herbal remedies will be needed. Plant substances, which are available as food and drink (e.g., herbal teas), can, however, be freely sold in the United Kingdom.

5.71 UNITED REPUBLIC OF TANZANIA

A national policy was issued in 2000, and a national office was established in 1989, as the traditional medicines section of the Department of Curative Services. Herbal medicines were not regulated in the United Republic of Tanzania. They had no regulatory status. There was no national pharmacopoeia or monographs available. In addition to this, there were no manufacturing requirements and no registration system for herbal medicines. A PMS system is being planned. There was no restriction on the sale of herbal medicines in the United Republic of Tanzania (WHO 2005). The Tanzania Food, Drugs and Cosmetics Act (2003) regulates both conventional and herbal drugs.

5.72 UNITED STATES

In the United States, the Food and Drug Administration (FDA) is the government agency that is primarily responsible for regulating foods and medicines. It is also responsible for regulating dietary supplements, such as herbal medicines, under the Dietary Supplement Health and Education Act (DSHEA) of 1994 (Abbott 2014). There was minimal control over the marketing of products from medicinal plants. A plant substance, whether listed in the pharmacopoeia or not, could be marketed in the country from the FDA. The system worked well because it was used with care largely by Hahnemann school of homeopaths. The situation changed with large-scale production of medicinal plants as therapeutic agents, more widespread use of homeopathic or OTC medicines, and their promotion. The term complementary or alternative medicines (CAMs) is more commonly used for traditional medicine in the United States. Integrative medicine refers to a practice that combines both conventional and CAM treatments for which there is evidence for safety and efficacy. Ang-Lee et al. (2001) discussed the history and developments of regulatory aspects of herbal remedies in detail. These were classified as dietary supplements in the Dietary Supplements, Health and Education Act of 1994. These laws exempted herbal medicines from the safety and efficacy regulation that prescription and OTC drugs had to fulfill (i.e., preclinical animal studies, premarketing controlled clinical trials, or PMS). The burden shifted to the FDA to show that a product is unsafe before it can be removed from the market. In addition, the inability to patent herbal medication discourages the manufacturer from performing the costly research and development (R&D) that is required for conventional drugs. It was felt that the regulations governing the use and marketing of herbal medicines were inadequate. Several simple changes could dramatically improve the appropriate use of herbal products. The creation of national standards for the constituents of specific herbs and greater incentives for R&D of study designs that reduce costs and study duration were suggested (Bent 2008). These should be more explicit and stringent. Therefore, the FDA classified CAM products into different categories: cosmetic, device, dietary supplement or food additive, drug, and new drug in its draft guidance for industries. These guidelines were later withdrawn following recommendations from several agencies including the American Herbal Product Association. FDA-regulated product categories under the Federal Food, Drug, and Cosmetic Act ("the Act") or Public Health Service Act ("PHS Act") include cosmetics, devices, dietary supplements, drugs, food, and food additives. A botanical product (herbal medicine or Chinese medicine) is subjected to regulation as biological product, cosmetic, drug, device, or food under "the Act" or "PHS Act." Many botanicals are used in cosmetics, for example, aloe vera, green tea; some Chinese or TCM herbs are available as regular food or spice, for example, ginger, star anise, ginseng, and so on. (Fan et al. 2012). The DSHEA specifies that supplements are to be regulated as foods rather than drugs, and thus dietary supplements could be sold freely in the market without proving evidence of efficacy or safety (Abbott 2014). They only comply with the labeling requirements for dietary supplements. The disease claims from traditional uses are, thus, not permitted on the label of the products. The DSHEA did not originally require that manufacturers report adverse effects to the FDA until 2006 when dietary supplement manufacturers have been required to report any "serious" adverse effects within 15 days of knowledge of the event. In addition, rules for GMPs were adopted in 2007. These rules set requirements for domestically marketed herbs that include meeting specifications for identity, purity, strength, and composition (Abbott 2014).

TCM products and raw herbs, however, can be further developed as new drugs through investigational new drug or new drug application (NDA) processes (Fan et al. 2012). The Center for Drug Evaluation and Research and New Guidance for Botanical Products differentiate between OTC and NDA. The document provides guidelines for the NDA route, including discussions on lesser demand for certain information for botanicals with an established history of use of waivers for related studies, or bridging studies may be considered by FDA on a case-to-case basis. Botanical products are finished, labeled products that contain vegetable matter as ingredients. Such products can be marketed in the United States with (1) OTC and (2) approved NDA product monographs. The manufacturer has to submit fresh application for adding a new ingredient or indication for amendment in the monographs. Eligibility to market the products is five years (for a new chemical entity) and three years (in the absence of patent protection) from the date of approval. If the product qualifies as NDA during the period of exclusivity, the FDA will not approve (or in some cases review) certain competitor products unless the second sponsor conducts all studies to demonstrate the safety and efficacy with preclinical pharmacology and toxicology studies (Verma 2013). Now, most practitioners are moving away from traditional knowledge to evidence-based medicines. The current U.S. regulatory mechanisms provide little assistance to ensure that commercial herbal preparations have predictable pharmacological effects, and the product labels provide accurate information. The potency of herbal medicines varies between different manufacturers of the same product and in different lots that are released into the market by the same manufacturer. Some companies have tried to standardize their products to a fixed concentration of the selected chemical constituents. The benefit of this effort is uncertain because most herbal products achieve their intended effect through combined synergistic actions of different ingredients or compounds.

5.73 UZBEKISTAN

The regulation of herbal medicines began in 1997; it is governed by the same laws and regulations as for conventional pharmaceuticals. Herbal medicines are regulated as prescription and OTC medicines. By law, medical, health, nutrient content, and structure or function claims may be made. A national pharmacopoeia and national monograph were in development. The regulatory requirement for the manufacture of the herbal medicines is limited to adhere to information in pharmacopoeias and monographs. No specific information was available about the control mechanism that is in use to enforce these requirements. The safety requirements for herbal medicines were the same as those that are for conventional pharmaceuticals; pharmaceutical inspection of laboratories for quality control was used to ensure the implementation of these requirements. No PMS system was established. Herbal medicines were sold in pharmacies as prescription and OTC medicines by licensed practitioners and without restriction (WHO 2005).

5.74 ZAMBIA

The national policy on TM/CAM, which was part of the national drug policy, was approved in 1997. While Zambia was not regulating herbal medicines, a bill was proposed for the same. Herbal medicines were sold with claims, including medical, health nutrient content, and structure or function claims; however, as the regulations were pending, these were not legally recognized. There was no national pharmacopeia or national monograph, no manufacturing or safety assessment regulatory requirements, no registration system, and no PMS system for herbal medicines. There was no restriction on the sale of the herbal medicines in Zambia (WHO 2005). The Pharmaceutical Act No. 14 of 2004 requires that products intended to be marketed in Zambia meet appropriate standards of good quality, safety, and efficacy and that they should be manufactured according to GMP requirements. One of the means for ensuring that herbal medicinal products meet the required standards of good quality, safety, and efficacy is by conducting product-specific premarketing assessments to determine whether the product should be registered. These guidelines developed by the Pharmaceutical Regulatory Authority have been prepared to provide information to applicants who intend to register herbal medicinal products in Zambia (Guidelines on Application of Registration of Herbal Medicines, http://www.zamra.co.zm/guidelines, accessed on May 10, 2015).

REFERENCES

Abbott, R. 2014. Documenting traditional medical knowledge. World Intellectual Property Organization (WIPO), March. Available at http://www.wipo.int, accessed on May 3, 2015.

Ang-Lee, M.K., Moss, J., and Yuan, C.S. 2001. Herbal medicines and perioperative care. *JAMA* 286: 208–216.

Bandaranayake, W.M. 2006. Quality control, screening, toxicity and regulation of herbal drugs. In: Ahmad, I., Aqil, F., and Owais, M. (eds.) *Modern Phytomedicine. Turning Medicinal Plants into Drugs*, Wiley-VCH Verlag GmbH & Co., Weinheim.

Bent, S. 2008. Herbal medicine in the United States: Review of efficacy, safety and regulation. *J Gen Intern Med* 23: 854–859.

Carvalho, A.C.B., Ramalho, L.S., Marques, R.F., and Perfeito, J.P.S. 2014. Regulation of herbal medicines in Brazil. *J Ethnopharmacol* 158: 503–506.

Castot, A., Diezzar, S., Deleau, N., Guillot, B., and Efthymiou, M.L. 1997. Pharmacovigilance off the beaten track: Herbal surveillance or pharmacovigilance of medicinal plants. *Therapie* 52: 97–103.

Chalut, D. 1999. Toxicological risks of herbal remedies. *Paediatr Child Health* 4: 536–538.

Chaudhury, R.R. 1992. *Herbal Medicine for Human Health*. World Health Organization, Regional Office for South East Asia, New Delhi.

Chaudhury, R.R. and Rafei, U.M. (eds.). 2002. *Traditional Medicine in Asia*. World Health Organization, Regional Office for South-East Asia, New Delhi.

Czerw, A. and Bilinska, M. 2013. Distribution of medicinal products in Poland: Light of legislative regulations. *J Stock Forex Trad* 3: 111.

Drugs and Cosmetics Act 1940 and Rules 1945, amendments up to 2005, Ministry of Health and Family Welfare, Govt. of India, New Delhi.

Drugs and Cosmetics Act 2010, 6th amendment rules, Department of Ayurveda, Yoga, Unani, Siddha and Homeopathy (AYUSH), Ministry of Health and Family Welfare, Govt. of India, New Delhi.

Drugs and Magic Remedies (Objectionable Advertisements) Act, 1954. Government of India.

Ekor, M. 2013. The growing use of herbal medicines: Issues relating to adverse reactions and challenges in monitoring safety. *Frontiers in Pharmacol* 4: 1–10.

Fan, T.P., Deal, G., Koo, H.L., Rees, D., Sun, H., Chen, S., Dou, J.H., Makrov, V.G., Pozharitskava, O.N., Shikov, A.N., Kim, Y.S., Huang, Y.T., Chang, Y.S., Jia, W., Dias, A., Wong, V.C., and Chan, K. 2012. Future development of global regulation of Chinese herbal products. *J Ethnopharmacol* 140: 568–586.

Gabriella, H., Fonnebo, V., Folkenberg, T., Hok, J., and Wiesener, S. 2013. Legal status and regulation of complementary and alternative medicine in Europe. *Lege Artis Medicinae* 23: 350–359.

GCP guidelines. 2001. *Good Clinical Practices for Clinical Research in India*, Chapter 7: Special concerns, Section 7.5: Clinical trials for herbal remedies and medicinal plants, pp. 53–55, Central Drugs Standards Control Organization, Ministry of Health and Family Welfare, Govt. of India, New Delhi.

George, P. 2011. Concerns regarding the safety and toxicity of medicinal plants: An overview. *J Appl Pharmaceut Sci* 1: 40–44.

"Guidelines for regulation of traditional/herbal medicines (local) in Uganda". 2009. Issued by National Drug Authority (NDA), Uganda. Available at http://www.nda.or.ug/docs/reg10.pdf.

Guidelines on application of registration of herbal medicines, issued by Pharmaceutical Regulatory Authority (PRA), Zambia. Available at http://www.who.int/medicines/areas/coordination/zambia_registration_herbal.pdf, accessed on May 10, 2015.

Homma, M., Oka, K., Nitisuma, J., and Itoh, H. 1993. Pharmacokinetic evaluation of traditional Chinese herbal remedies. *Lancet* 341: 1595.

Human Medicine Regulations. 2012. Part 7, Regulations 125, 241, United Kingdom.

Katoch, V.M. and Kumar, V. 2012. Head Division of Basic Medical Sciences, Indian Council of Medical Research, New Delhi, personal communications, September 12 and December 26.

Maegawa, H., Nakamura, T., and Saito, K. 2014. Regulation of traditional herbal medicinal products in Japan. *J Ethnopharmacol* 158: 511–515.

MHRA. 2013. End of transitional arrangements for the directive on traditional herbal medicinal products (Directive 2004/24/EC, amending directive 2001/83/EC), November, MHRA, United Kingdom.

Mukherjee, P.K., Venkatesh, M., and Kumar, V. 2007. An overview of the development in regulation and control of medicinal and aromatic plants in the Indian system of medicine. *Boletin Latinoam Caribe Plant Med Aromaticas* 6: 129–137.

Musgrave, I. 1998. Herbal medicines: Toxic side effects, drug interactions and regulations. Available at http://theconversation.com/herbal-medicines-toxic-side-effects-and-drug-interactions-23118, accessed March 7, 2014.

Osuide, G.E. 2002. Regulation of herbal medicines in Nigeria: The role of the National Agency for Food and Drug Administration and Control (NAFDAC). In: Iwu and Wootton (eds.) *Ethnomedicine and Drug Discovery* 1: 249–258, Elsevier Science BV, Amsterdam, The Netherlands.

Pelkonen, O., Pasanen, M., Lindon, J.C., Chan, K., Zhao, L., Deal, G., Xu, Q., and Fan, T. 2012. Omics and its potential impact on R&D and regulation of complex herbal products. *J Ethnopharmacol* 140: 587–593.

Peschel, W. 2007. *The Impact of the European Traditional Use Directive on the Herbal Product Markets in the United Kingdom, Germany and Spain*. Available at http://www.uv.es/prietojm/.../Peschel_b_BLACPMA_V6_N4.pdf, accessed on May 5, 2015.

Peschel, W. 2014. The use of community herbal monographs to facilitate registrations and authorisations of herbal medicinal products in the European Union 2004–2012. *J Ethnopharmacol* 158: 471–486.

Rates, S.M.K. 2001. Review: Plants as source of drugs. *Toxicon* 39: 603–613.
Regulatory status of herbal medicines in Iran. Available at http://www.itmrc.org/regulator1.htm, accessed on May 4, 2015.
Rivera, J.D., Loya, A.M., and Ceballos, R. 2013. Use of herbal medicines and implications for conventional drug therapy medical sciences. *Altern Integr Med* 2: 130.
Sahoo, N. and Manchikanti, P. 2013. Herbal drug regulation and commercialization: An Indian Industry perspective. *J Altern Complement Med* 19: 957–963.
Saper, R.B., Kales, S.N., Paquin, J., Burns, M.J., Eirenberg, D.M., Davis, R.B., and Phillips, R.S. 2004. Heavy metal content of Ayurvedic herbal medicine products. *JAMA* 292: 2868–2873.
Sensho, Y. 2014. Japanese Embassy at New Delhi, India, personal communication, March.
Shikov, A.N., Pozharitzkava, O.N., and Makarov, V.G. 2012. Regulation of herbal medicinal products in Russia. *Planta Medica* 78-OP10.
Smith, A., Jogalekar, A., and Gibson, A. 2014. Regulation of natural health products in Canada. *J Ethnopharmacol* 158: 507–510.
Symposium on Strategies for Safety Study Requirements for Herbal Formulations and 30th Annual Conference of Society of Toxicology. 2010. December 9–11, Jamia Hamdard, New Delhi.
Thailand Drug Act BE 2510. 1967. Available at http://www.globinmed.com/index.php?option=com_content&view=article&id=104195:thailand-drug-act-b-e-2510-1967&catid=259, accessed on May 8, 2015.
UNESCO. 2013. *Report of the International Bioethics Committee on Traditional Medicine Systems and Their Ethical Implications. SHS/EGC/IBC-* 19/12/3 Rev. Paris, February 8.
van Galen, E. 2014. Traditional herbal medicines worldwide, from reappraisal to assessment, in Europe. *J Ethnopharmacol* 158: 498–502.
Verma, N. 2013. Herbal medicines: Regulation and practice in Europe, United States and India. *Int J Herb Med* 1: 1–5.
WHO. 1998. Regulatory situation of herbal drugs: A worldwide review. World Health Organization, Geneva.
WHO. 2004. *Guidelines on Safety Monitoring of Herbal Medicines in Pharmacovigilance Systems*. World Health Organization, Geneva.
WHO. 2005. National policy on traditional medicine and regulation of herbal medicines: Report of a WHO global survey. World Health Organization, Geneva.
WHO. 2006–2014. *International Regulatory Cooperation for Herbal Medicines (IRCH)*. World Health Organization, Geneva.
Willcox, M., Sanogo, R., Diakite, C., Giani, S., Paulsen, B.S., and Diallo, D. 2012. Improved traditional medicines in Mali. *J Alternative Complementary Med* 18: 212–220.
Zeman, S. 2014. Australian High Commission at New Delhi, India, personal communication, April.

6 A Critical Appraisal of the Available Data

This review embodies a survey report on more than 1500 medicinal plants and herbal products. More than 100 examples of toxic or adverse side effects or herb + herb and herb + modern medicine interactions have been cited. The list is only illustrative and not exhaustive. It includes 59 examples from 27 countries (Argentina, Australia, Austria, Bangladesh, Belgium, Brazil, Canada, China, France, Germany, Ghana, Hong Kong, Iran, Italy, Japan, Malaysia, Mauritius, Morocco, the Netherlands, Nigeria, Portugal, South Africa, Spain, Sri Lanka, Turkey, the United Kingdom, and the United States), of the effects of 75 commonly used medicinal plants and 12 herbal formulations. The adverse effects range from mild gastrointestinal problems (nausea, vomiting, diarrhea) and allergic skin reactions (inflammation, rash, itching) to severe toxic effects on the liver; kidneys; cardiovascular, respiratory, and nervous systems; reproductive function; and mutagenic, teratogenic, and carcinogenic effects. Some of these effects were serious enough to necessitate hospitalization and intensive medical care. Early detection and proper diagnosis by alert physicians helped in the survival of affected persons, but some cases of permanent disability and death have also been reported. Some striking reports are as follows.

a. *Hussin (2001) cited a report by the Ministry of Health, Malaysia.* It indicated that about 37% of 5000 renal problems in the country might be attributed to the chronic use of traditional herbal preparations.
b. *A study by Hailer et al. (2002) stated that diagnosis of herbal-related toxic hepatitis may be hindered by patients' reluctance to admit the use or disclose the names of alternative medicines.* A survey revealed that only 40% of the patients in the United States informed their primary healthcare physician about it, resulting in failure to recognize or chemically identify hepatotoxic herbs. The figure may be much lower as mostly such drugs are self-prescribed and taken concurrently with modern medicines. The researcher demonstrated the importance of early diagnosis of herbal drug usage by describing two contrasting case studies. Early diagnosis by the alert physician, patient cooperation, and systematic hospital care in the first case lead to quick recovery. Poor history, multiple product use, self-help, and reluctance to seek medical advice in the other patient resulted in serious disease necessitating liver transplantation, which could have been avoided.
c. *Plants containing coumarin derivatives (e.g., Panax species) and Ginkgo biloba may lead to hemorrhages by their own chronic use or through enhancing the effects of anticoagulants (e.g., Dicumarol)* (Ang-Lee et al. 2001; Rates 2001).

d. *An extensive review of nephrotoxicity due to alternative medicines by Luyckx (2012) tabulating 56 remedies (herbal: 43, mineral or animal origin: 13) from 35 countries.* The author stressed that kidneys are particularly vulnerable to toxic effects from such drugs. A prominent example cited was aristolochic acid-induced nephropathy. This developed into a small outbreak in Belgium and then a pathophysiological mechanism for a previously unexplained endemic disease and later as a significant risk factor for chronic kidney disease (CKD) in population-based studies.
e. *A study by Abraham et al. (2010) from Germany on hepatotoxic and possibly the mutagenic or carcinogenic effects of coumarins.* Mutagenicity or carcinogenicity was discounted by the European Food Safety Authority. Nutritional exposure to coumarins is mainly due to Cinnamon (*Cassia cinnamon*), which is used as a flavoring agent in foods (tea, cereals, cookies or almond cookies, desserts, chocolates, and wine). Its consumption is considerably increased during the Christmas season.
f. *Unintentional (wrong botanical identification) and intentional (adulteration with cheaper substitutes) use of wrong herbal remedy may result in mild-to-serious toxic or adverse effects, e.g., coffee substitutes causing such problems in Brazil* (Rates 2001).

Risk groups for herbal drug toxicity or herbal drug + medicine interactions include pregnant and nursing women, infants, children, elderly persons, patients with diseases (such as diabetes, cardiovascular, hepatic, respiratory, and neuropsychiatric disorders), pre- and postoperative cases under other life-sustaining medications (McGregor et al. 1989; Ang-Lee et al. 2001; Hussin 2001; Hailer et al. 2002). Unfortunately, studies on this aspect are scarcely conducted; the increasing use of herbal drugs globally means that the available reports are only a tip of the iceberg. A study by Ang-Lee et al. (2001) on herbal medicines and perioperative care is, particularly, noteworthy. More investigations are warranted to unearth the magnitude and hazards of identifications between herbal products and conventional medicines. A case report on a 51-year-old man diagnosed with acute kidney injury with toxic encephalopathy following consumption of herbal juice from the raw tubers of *Dioscorea quinqueloba* was recently reported. Such serious adverse effect-causing medicines need to be classified as prescription drug requiring stringent regulation (Kang and Heo 2015). A recent research on PubMed revealed that at least 65 commonly used herbs, herbal drugs, and supplements and 111 herbs or herbal mixtures of traditional Chinese medicine have been reported to cause hepatotoxicity or liver disease, necessitates stricter regulatory surveillance for herbal drugs (Teschke and Eickhoff 2015).

Does this mean that all herbal drugs are useless and harmful? The answer is "No"; the question rather raises many other questions. Is it genuine concern for herbal drug safety or overreaction due to their widespread or increasing worldwide use affecting the sales of modern medicines? Are the reasons governed by economic or business considerations or truly scientific? It has been argued that herbal drugs (e.g., Ayurvedic) have been in use for more than 4000 years, and a large proportion of the Afro–Asian population depends on traditional medicines.

Several toxic plants, used in traditional medicine, were found to be useful for humanity for the treatment of serious diseases, for example, *Colchicum autumnale* for congestive heart failure, Digitalis for cardiac insufficiencies, *Citrullus colocynthus* for diabetes, and so on (Bnouham et al. 2006). A report by Katiyar (2012) on the safety evaluation of Ayurvedic drugs gives some convincing arguments considering the number of Ayurvedic practitioners in India to be approximately 300,000 and assuming the number of patients receiving Ayurvedic medicines to be 10 per day (4 old and 6 new cases) by a physician, which amounts to be approximately 3 million patients per day. The researcher stated that safety evidence is being generated on commonly prescribed Ayurvedic herbal products almost on a daily basis. Further, if the herbal medicines are so harmful,

a. Why is our herbal wealth being indiscriminately depleted by increasing exports of crude medicinal plants?
b. Why are repeated attempts being made to patent these drugs?

The cases of *Curcuma longa* (Haldi or Turmeric) and *Ocimum sanctum* (Tulsi or Basil) are examples. According to a recent news proper report (*Hindustan Times* 2013), the U.S. multinational company MetaProteomics filed a patent application at the Canada Intellectual Property Office under the title "Curcuminoid composition exhibiting synergistic inhibition of the expression of Cycloxygenase-2 drugs for the treatment of inflammation, psoriasis and gastritis to be novel." The Traditional Knowledge Digital Library, a unit of Council of Scientific and Industrial Research, cited ancient Indian classical literature to counter this contention. Within a period of 87 weeks, a 10-year-old attempt to pirate Indian traditional knowledge was foiled by the government of India.

Polypharmacy is generally used in Ayurveda, Siddha, and Unani medicines. It is claimed that different ingredients of the formulation enhance the efficacy and reduce the toxicity of each other. Further, *Shodhna* and *Marna* (traditional purification procedures) detoxify the toxic effects of herbal drugs. Such procedures (e.g., treating with lemon juice or boiling with water) have not been scientifically validated (or refuted) by comparative studies on raw and purified herbal products.

Many studies report high metal content in herbal medicines (Saper et al. 2004; Symposium 2010). Traditional systems of medicine, practiced on the Indian subcontinent, also use drugs of mineral origin including heavy metals. One of the authors (SBV) has been involved in scientific investigations on calcined mineral preparations: *Bhasmas* and *Kushtas* (used in Ayurveda and Unani-Tibb) for more than a decade. The studies were carried out under projects from the Indian Council of Medical Research and Central Council for Research in Unani Medicine. Comparative studies on gold preparations—*Swarna Bhasma* (Ayurveda), *Kushta Tila Kalan* (Unani-Tibb), and Auranofin (modern medicine)—revealed interesting results. The latter is known to be an immunosuppressant and used for the treatment of rheumatoid arthritis. Chrysotherapy is associated with several problems (not effective in all cases: 20% of patients get no benefit, low therapeutic index: 25%–40% of patients develop toxic reactions necessitating discontinuance of therapy in half of such cases, high relapse rate, etc.); calcined preparations used in Indian systems of medicine, in contrast, exhibited immunostimulant activity (Bajaj et al. 1999). These preparations also showed analgesic, anticataleptic, antianxiety, antidepressant, glycogen-sparing or antifatigue, antioxidant, nootropic, and neuroprotective effects against global and focal models of ischemia in experimental animals. The products attenuated stress-induced alterations in the brain catecholamines, serotonin, and plasma cortisol levels. While these effects were observed in doses up to 25 mg kg^{-1}, the maximum tolerated dose was found to be more than 2 kg^{-1}. The preparations showed a wide margin of safety. No weight loss or untoward effect on a battery of behavioral parameters was observed. When given in diet to young rat pups for six to eight weeks, the treated animals showed an increase in body weight or growth rate versus controls. Chronic toxicity studies were not done and are needed. The mere presence of a metal in a herbal product is not a proof of its toxicity. The forms of metal and exposure levels are required to be considered (Bajaj and Vohora 1998, 1999, 2000; Shah and Vohora 2002; Shah et al. 2005; Vohora and Athar 2007).

A perusal of the reports on formulations indicates that these are relatively safe or show toxic effects that are not of serious nature. Does it prove the belief that individual components play a corrective role? The number and standard of available studies are not adequate for any meaningful conclusions.

Are herbal drugs cheap? The present review is not addressing this question. Herbal drugs are promoted as relatively affordable versus modern medicines. Hussin (2001) remarked that this was probably true earlier but is not so now. These products are fast becoming expensive (sometimes more costly versus conventional medicines). Increased demand or fashion for herbal drugs and cosmetics, recent spurt in online sales, and aggressive marketing are the factors that are responsible for such rise in prices.

The present status of regulatory laws in 73 countries (Angola, Antigua and Barbuda, Argentina, Armenia, Australia, Austria, Azerbaijan, Bangladesh, Belarus, Belgium, Benin, Bhutan, Botswana, Brazil, Bulgaria, Burkina Faso, Burundi, Cameroon, Canada, Central African Republic, Chad, China, Comoros, Congo, Czech Republic, Denmark, Estonia, Ethiopia, Equatorial Guinea, Finland, France, Georgia, Germany, Ghana, Greece, Guinea, Hungary, Iceland, India, Indonesia, Iran, Ireland, Israel, Japan, Kenya, Korea, Maldives, Mali, Mauritius, Myanmar, Nepal, Netherlands, Nigeria, Norway, Poland, Portugal, Romania, Russia, South Africa, Spain, Sri Lanka, Sweden, Switzerland, Tajikistan, Thailand, Turkey, Uganda, Ukraine, the United Kingdom, United Republic of Tanzania, the United States, Uzbekistan, and Zambia) revealed that these are mostly vague or weak and not stringent enough to enforce existing laws and do not ensure the safety of herbal products. A WHO document (2004) laments that about 90 countries (less than half of the member states) regulate herbal drugs, and even a smaller number have a requisite system in place for regulation or qualifications of providers of such products.

In India, time-to-time amendments of the Drugs and Cosmetics Act and Rules (1940) and Drugs and Cosmetics Rules (1945) are a good sign. The government of India has made heavy metal testing mandatory after the publication of Saper et al. (2004). The latter proved to be a blessing in disguise. It compelled the government to be more responsible. A general look at the overall status of regulatory aspects of herbal drugs worldwide indicates that in a majority of the countries, factors other than scientific (political, religious, and economic compulsions and vested business interests) control these issues at the cost of public health. The Drugs and Magic Remedies (Objectionable Advertisement) Act is in place but openly flouted on the Indian subcontinent. This is particularly true for herbal or herbomineral products claiming potentiation of male sexual dysfunction or impotency and infertility. Strict implantation of the regulations with deterrent punishment for the offenders is urgently needed.

We do not intend to say that all herbal drugs are harmful. Nearly three-fourths of the global population (particularly in Afro-Asian countries) depend on herbal or traditional drugs for the treatment of many diseases. Use of traditional herbal or herbomineral drugs dates back to 3000 BC. Data on safety aspects on these drugs, with experience of centuries, are being generated on an almost daily basis in human subjects. Studies on experimental animals and clinical trials in a limited number of patients cannot match it. Polypharmacy is the rule, rather than an exception, in traditional medicinal systems. It is claimed that the different ingredients in a formulation enhance the efficacy and reduce the toxicity of each other. These claims, in most cases, have not been scientifically validated (or refuted).

Recently, a very disturbing newspaper report (*Hindustan Times* 2013) indicated that blind faith in alternative therapy delays cancer treatment. Almost 70% of new cancer cases reported to the B.R. Ambedkar Institute Rotary Cancer Hospital, All India Institute of Medical Sciences (AIIMS), New Delhi, are beyond the critical stages, when the disease cannot be treated easily. Dr. P.K. Julka, Head, Department of Radiation Oncology, AIIMS, stated that this adversely affects the outcome of the treatment. Other cancer hospitals face similar problems. The case of a 48-year-old farmer undergoing treatment in West Delhi Action Cancer Hospital illustrates the point. The patient was diagnosed with cancer nine months before admission and was advised to undergo surgery for breast cancer followed by chemotherapy. Instead, she went to a traditional medicine practitioner who charged the patient Rs 35,000 for drugs with the promise that they will cure her. She reported at stage 4 of breast cancer and is undergoing treatment now after progression to disease at such an advanced stage. Prognosis in such cases is poor as valuable time is wasted on drugs with unknown effects that are sold to gullible persons with false claims.

It is concluded that no effective drug (traditional or modern) is totally good or bad. A balanced dispassionate approach is warranted. Herbal drugs (single medicinal plants or formulations) should be treated with the same degree of respect and responsibility as for conventional medicines. Problems of indiscriminate use, self-prescription, concurrent use with modern medicines without informing the physician or hospital, administration of wrong herbal drug (incorrect botanical identification), quackery, adulteration, and so on, need to be curbed with firm resolve, strict enforcement of regulatory laws with deterrent punishment by the government, and public cooperation. Indian systems of medicine (ISMs) have many useful remedies, but the most innocuous herbal drugs have also caused toxic or adverse effects and hazardous drug interactions by their improper use. Adamant or blind faith in ancient classical texts has hampered progress in ISM and has not allowed them to keep pace with time, the latest research, and advancements and flourish to their full potential. Paradoxically, this has been done by exponents of ISM who intend to promote traditional medicine. These issues need to be tackled urgently before they reach an alarming stage.

REFERENCES

Abraham, K., Wohrlin, F., Lindtner, O., Heinmeyer, G. and Lampen, A. 2010. Toxicology and risk assessment of coumarin: Focus on human data. *Mol Nutr Food Res* 54: 228–238.

Ang-Lee, M.K., Moss, J., and Yuan, C.S. 2001. Herbal medicines and perioperative care. *JAMA* 286: 208–216.

Bajaj, S. and Vohora, S.B. 1998. Analgesic activity of Unani gold preparations used in Indian system of medicine. *Indian J Med Res* 108: 151–162.

Bajaj, S. and Vohora, S.B. 1999. Anti-cataleptic, anti-anxiety and anti-depressant activity of gold preparations used in Indian system of medicine. *International Congress on Frontiers of Pharmacology and Therapeutics in 21st Century*, New Delhi.

Bajaj, S. and Vohora, S.B. 2000. Anti-cataleptic anti-anxiety and anti-depressant activity of gold preparations used in Indian system of medicine. *Indian J Physiol Pharmacol* 32: 339–346.

Bajaj, S., Ahmad, I., Fatima, M., Raisuddin, S. and Vohora, S.B. 1999. Immunomodulatory activity of Unani gold preparations used in Indian system of medicine. *Immunopharmacol Immunotoxicol* 22: 151–162.

Bnouham, M., Mehrfour, F.Z., Elachoui, M., Legssyer, A., Mekhfi, H., Lamnouer, D., and Ziyyat, A. 2006. Toxic effects of some plants used in Moroccan traditional medicine. *Moroccan J Biol* 2–3: 21–30.

Drugs and Cosmetics Act 1940 and Rules 1945, Ministry of Health and Family Welfare, Govt. of India, New Delhi.

Hailer, C.A., Dyer, J.E., Ko, R.J., and Olson, K.R. 2002. Making a diagnosis of herbal-related toxic hepatitis. *West J Med* 176: 39–44.

Hindustan Times. 2013. Faith in alternative therapy delays cancer treatment. New Delhi, February.

Hussin, A.H. 2001. Adverse effects of herbs and drug-herbal interactions. *Malaysian J Pharmacy* 1: 39–44.

Kang, K.S. and Heo, S.T. 2015. A case of life-threatening acute kidney injury with toxic encephalopathy caused by *Dioscorea quinqueloba*. *Yonsei Med J* 56: 304–306.

Katiyar, C.K. 2012. Safety evaluation of Ayurvedic drugs: Regulatory requirements, issues and perspectives. *Herbal Products: Regulatory Aspects*, T.N. Medical College, Mumbai.

Luyckx, V.A. 2012. Nephrotoxicity of alternative medicine practice. *Adv Chronic Kidney D* 19: 129–141.

McGregor, F.B., Abenerthy, V.E., Dahabra, S., Cobden, I., and Hayes, P.C. 1989. Hepatotoxicity of herbal remedies. *BMJ* 299: 1156–1157.

Rates, S.M.K. 2001. Review: Plants as source of drugs. *Toxicon* 39: 603–613.

Saper, R.B., Kales, S.N., Paquin, J., Burns, M.J., Eisenberg, D.M., Davis, R.B., and Phillips, R.S. 2004. Heavy metal content of Ayurvedic herbal medicine products. *JAMA* 292: 2868–2873.

Shah, Z.A. and Vohora, S.B. 2002. Anti-oxidant/restorative effects of calcined gold preparations used in Indian systems of medicine against global and focal models of ischemia. *Pharmacol Toxicol* 99: 254–259.

Shah, Z.A., Sharma, P., and Vohora, S.B. 2005. Attenuation of stress-elicited brain catecholamines, serotonin and plasma corticosterone levels by calcined gold preparations used in Indian system of medicine. *Pharmacol Toxicol* 96: 469–474.

Symposium on Strategies for Safety Study Requirements for Herbal Formulations and 30th Annual Conference of Society of Toxicology. 2010. December 9–11, Jamia Hamdard, New Delhi.

Teschke, R. and Eickhoff, A. 2015. Herbal hepatotoxicity in traditional and modern medicine: Actual key issues and new encouraging steps. *Frontiers in Pharmacol* 6: 72.

Vohora, S.B. and Athar, M. 2007. *Mineral Drugs Used in Ayurveda and Unani Medicine*. Narora Publishing House Pvt Ltd, New Delhi.

WHO. 2004. *Guidelines on Safety Monitoring of Herbal Medicines in Pharmacovigilance Systems*. World Health Organization, Geneva.

Index

Page numbers followed by t indicate tables.

S